Applications of Remote Sensing

Edited by Sayed Hemeda

Published in London, United Kingdom

Applications of Remote Sensing
http://dx.doi.org/10.5772/intechopen.1001531
Edited by Sayed Hemeda

Contributors
Sayed Hemeda, Nisha Rani, Sumit Kumar Ahirwar, V.M. Anoop, K.V. Krishnamurthy, Maan Habib, Amanda Fruehauf, Adriana Afonso Sandre, Magda Lombardo, Paulo Renato Mesquita Pellegrino, Nurhussen Ahmed, Worku Zewdie, Daniel Devote Basil, Lawrence Hart, Gbenga Ajayi

First published in London, United Kingdom, 2023 by IntechOpen
IntechOpen is the global imprint of INTECHOPEN LIMITED, registered in England and Wales, registration number: 11086078, 5 Princes Gate Court, London, SW7 2QJ, United Kingdom

British Library Cataloguing-in-Publication Data
A catalogue record for this book is available from the British Library

Additional hard and PDF copies can be obtained from orders@intechopen.com

Applications of Remote Sensing
Edited by Sayed Hemeda
p. cm.
Print ISBN 978-1-83769-634-5
Online ISBN 978-1-83769-633-8
eBook (PDF) ISBN 978-1-83769-635-2

Meet the editor

Professor Sayed Hemeda obtained his Ph.D. in Civil Engineering from Aristotle University of Thessaloniki, Greece. He is the manager of Heritage Science Programs at the Liberal Art and Culture Centre (LACC), and a professor at the Basic and Applied Science Institute (BAS), Egypt-Japan University of Science and Technology (E-JUST). He is the first Professor of Heritage Science and Architectural Preservation of Architectural Heritage, Conservation Department, Faculty of Archaeology, Cairo University, Egypt. He is also the former manager of the Historic Buildings Conservation Center, Cairo University. He is the recipient of many awards from Cairo University, including prizes for scientific excellence (2017), Prize of Scientific encourage (2014), and best Ph.D. thesis (2009-2010). He was also awarded the 2019 Union of Arab Archaeologists Prize for Academic Excellence. He has published eighty-five articles and thirty-three books. He has given more than fifty-eight invited lectures in sixteen countries. His primary interests are geotechnical engineering for architectural heritage preservation and engineering data analysis including pattern recognition as applied to primarily analytical data from various sources such as objects of cultural significance. He is editor-in-chief of the *International Journal of Advances in Geological and Geotechnical Research*. He is also an editorial board member for many organizations and publications including publishers like Springer Nature, IntechOpen, and Bentham, and journals such as *Open Journal of Geology, Progress of Electrical and Electronic Engineering*, and *Geoscience Journal*. Dr. Hemeda is a scientific and organization committee member for many international conferences.

Contents

Preface

This book focuses on new perspectives and advanced theoretical and application research in advances in remote sensing and geodes. It is concerned with the study of scientific problems of remote sensing and geodesy and related interdisciplinary sciences. It will be of interest to surveyors, geodesists, and geophysicists, as well as practical space science and civil and architectural engineers. The book discusses some contemporary issues related to remote sensing and geodesy in engineering projects, which are critical components in civil construction and often require detailed management techniques and unique solutions to address failures and remedial measures.

This book discusses a broad range of topics in remote sensing and geodetic research, including surveying engineering, numerical methods, and software developments in remote sensing and geodetic networks, deformation analysis, adjustment theory and application of mathematical statistics, satellite geodesy, physical geodesy, geophysical geodesy, geodynamics, geometric geodesy, Global Navigation Satellite Systems (GNSS), and satellite gravimetry.

Chapter 1 is an introductory chapter that discusses the recent applications of remote sensing in the geoenvironmental hazard analysis and risk assessment for coastal historic towns and archaeological sites. The first case is the flash flooding of Derna city in Libya due to the Daniel hurricane and the second case is the rock-cut tombs in Valley of Kings in Luxor, Egypt.

Chapter 2, "Mineral Mapping through Advanced Airborne Hyperspectral Remote Sensing Techniques", discusses mineral mapping. In the mapped area, carbonate is present in the form of altered minerals in metabasalt, and iron leached surface indicates hematite mineral. The remote sensing outputs are also supported by the petrographic studies. The approach used in this study has great implications for mineral mapping.

Chapter 3, "Geodetic Data Processing Using Soft Computing Techniques", discusses using soft computing techniques in geodetic data processing and examines related challenges and future direction. Soft computing techniques, including artificial neural networks, fuzzy logic, and evolutionary algorithms, are used more frequently in geodetic data processing due to their ability to handle complex, imprecise, and uncertain data.

Chapter 4, "Urban Trees in the Metropolitan Region of São Paulo: A Study of Geodesign and Ecosystem Services", uses Geographical Information Systems (GIS) and a geodesign framework, along with the GISColab platform, to study urban afforestation in São Paulo. Thus, the work aims to seek nature-based solutions, such as increasing urban afforestation, to promote socio-environmental quality for the region.

Chapter 5, "Modeling Invasive *Prosopis juliflora* Distribution Using the Newly Launched Ethiopian Remote Sensing Satellite-1 (ETRSS-1) in the Lower Awash River

Basin, Ethiopia", discusses Ethiopia's successful launch of its first earth-observing satellite sensor in December 2019 for the purpose of managing natural resources and enhancing agriculture. This study aims to evaluate the potential of Ethiopian Remote Sensing Satellite 1 (ETRSS-1) to detect and map the Prosopis juliflora distribution for the first time. To better test its potential, a comparison was made between the novel Sentinel-2 Multispectral Instrument and Landsat-8 Operational Land Manager datasets.

Finally, Chapter 6, "Marine Geodesy for Hazard Investigation of EDOP Production Platform, Akwa Ibom State, Nigeria", focuses on the application of marine geodesy in evaluating the potential risks that affect the positioning of any marine vessel that is waiting for a safe anchorage for further instruction in Mobil Oil Mining Lease (OML) 112, about 44-km offshore of Akwa Ibom state, Nigeria.

I would like to express my gratitude to IntechOpen for their efforts in publishing this book and to the authors of the accepted manuscripts for their work and patience.

Sayed Hemeda
Professor,
Faculty of Archaeology,
Conservation Department,
Cairo University,
Giza, Egypt

Chapter 1

Introductory Chapter: Remote Sensing – Advances and Applications

Sayed Hemeda

1. Introduction

The term remote sensing was first used in the 1960s. Remote sensing is the science of obtaining information about objects or areas from a distance, usually from aircraft or satellites [1]. It is measuring or obtaining information about some characteristics of a phenomenon in a recording device that does not come into direct contact with the phenomenon that we are studying. It is the process of collecting data in the waves between the ultraviolet and the radio range.

2. Earth observing system (EOS)

The Earth Observing System (EOS) is a program of the United States National Aeronautics and Space Administration (NASA) consisting of a science component, a large advanced data system, and a space technology component consisting of a series of polar orbits and intermediate orbits. Tilt satellites for long-term observation of the Earth as an integrated long-term observing system include observations of the Earth's surface and features, biosphere, atmosphere, cryosphere, and oceans [2].

3. Advanced remote sensing

Advanced Remote Sensing is an application-based digital reference that provides a single source for the mathematical and physical concepts needed to collect and assimilate remote sensing data. It offers the latest advanced space technologies and technologies to estimate Earth's surface variables from a variety of data types, including optical and laser sensors such as RADAR and LIDAR.

Scientists in a number of different fields, including geography, geology, geophysics, atmospheric sciences, environmental sciences, planetary sciences, engineering, and ecology, have almost unlimited access to critically important data from remote sensing techniques, their developments, and applications [3].

4. The importance of remote sensing

All types of remote sensing: aerial photographs, satellite views, radar, LIDAR etc. are very important and provides abundant information about the Earth. They help to continuous surveillance of land and resources.

Examples of the importance of remote sensing:

- Study of sustainable natural resources
- Assessment of natural hazard
- Production of digital maps

5. Components of a remote sensing system

The remote sensing system that uses *electromagnetic radiation* consists of:

- Source: the source of electromagnetic radiation may be sunlight or heat.
- Interaction with Earth's surface phenomena: depends on the amount of reflected or transmitted radiation.
- Interaction with the atmosphere: where the energy passing through the atmosphere is affected.
- Sensors: recording radiation after its interaction with the Earth's surface and atmosphere.

Remote sensors can be either passive or active:

Passive sensors respond to external stimuli. They record natural energy that is reflected or emitted from the Earth's surface. The most common source of radiation detected by passive sensors is reflected sunlight.

In contrast, active sensors use internal stimuli to collect data about Earth. For example, a laser-beam remote sensing system projects a laser onto the surface of Earth and measures the time that it takes for the laser to reflect back to its sensor.

6. The importance of remote sensing in geography

- Monitoring the spatial distribution of terrestrial phenomena within a wide framework.
- Studying changing phenomena such as floods and traffic.
- Permanent recording of phenomena so that they can be studied at any time later.
- Recording data that the naked eye cannot see, as human eyes are sensitive to visible rays.

- Making quick and accurate measurements to a large extent for distances, areas, and heights.

7. Application of remote sensing

Remote sensing has a wide range of applications and benefits in many different fields and study areas:

- *Coastal applications:* Permanent, dynamic, and long-term monitoring of shoreline changes; tracking sediment transport; and mapping coastal features and descriptions. The data can be used in coastal surveying, mapping, and erosion prevention methods.

- *Ocean applications:* Monitoring of ocean circulation and current systems, measuring ocean temperature and wave height, and tracking sea ice. The data can be used to better understand the oceans and how to best manage ocean resources.

- *Natural disasters and hazard assessment: Tracking hurricanes, strong seismic events and earthquakes, erosion, and flooding.* Data can be used to assess the impacts of a natural disaster due to climate crises and create preparedness strategies to be used before and after a hazardous event.

- *Natural and cultural heritage preservation and management:* Monitoring land use, mapping wetlands and charting wildlife habitats. Data can be used to minimize the damage that urban growth has on the environment and archeological sites and help decide how to best protect natural and heritage resources.

  Egypt is particularly vulnerable to the impact of climate crises resulting from current and future climate change and variations. Climate change leads to an increase in the number and impact of natural disasters and hazards and their devastating impact on our natural and cultural heritage sites in Egypt.

  The increase in the number of risks and hazards related to climate change crises has exposed the cultural and natural heritage sites in Egypt to a greater danger than ever before, which imposes new challenges on conservation scientists and conservators to preserve the urban historic fabric and the built heritage in Egypt [4].

- *Water:* Egypt suffers from a shortage of water and difficulty in obtaining it. Remote sensing methods help in exploring groundwater locations and earthed and hidden water channels.

- *Minerals:* Sensors work in exploring mineral and petroleum ores.

- *Agriculture:* Inventorying agricultural crops and long-term crop monitoring and detecting plant diseases and moisture of the land.

- *Engineering works:* Used in studying construction and urban projects in addition to the risk assessment through the stages of construction.

8. Torrential rain wreaks havoc in Libya

Hurricanes, commonly known as tropical cyclones, are low-pressure systems with organized thunderstorm activity that form over tropical or subtropical waters.

Hurricanes are large, swirling storms. They produce winds of 119 kilometers per hour (74 mph) or higher. This is faster than a cheetah, the fastest animal on Earth. Winds from a hurricane can damage buildings and trees.

Hurricanes form over warm ocean waters. Sometimes they hit the ground. When a hurricane reaches land, it pushes a wall of ocean water onto the shore. This wall of water is called a storm surge. Heavy rains and storm surge from a hurricane and can cause flooding.

Once a hurricane forms, meteorologists predict its path. They also anticipate how powerful it will be. This information helps people prepare for the storm.

Scientists do not know exactly why or how a tornado forms. But they know that two main ingredients are needed.

One of the ingredients is warm water. Warm ocean water provides the energy a storm needs to develop into a hurricane. Normally, the surface water temperature must be 26°C (79°F) or higher for a hurricane to form.

The other element is the wind, which does not change much in speed or direction as it rises in the sky. Winds that change a lot with height can tear through storms.

Heavy rains from a storm called Storm Daniel hit parts of the central and eastern Mediterranean, leading to devastating floods and massive loss of life in Libya, the worst-hit country, as well as in Greece, Turkey, and Bulgaria.

The Libyan Red Crescent Society announced that up to 10,000 people are missing, according to the International Federation of Red Cross and Red Crescent Societies. The number of victims was not immediately confirmed, but it is feared that thousands may have died after the waters resulting from the collapse of the dams flooded the city of Derna.

The National Meteorological Center in Libya said that the storm reached its peak in northeastern Libya on September 10, with strong winds whose speed ranged between 70 and 80 km/h, which led to a communications outage and the fall of electric towers and trees. Heavy rains ranging from 150 to 240 mm caused flash floods in many cities, including Al Bayda, which recorded the highest daily rainfall of 414.1 mm (from September 10, 8 a.m. to September 11, 8 a.m.). The National Center of Meteorology said that this was a new record for rainfall.

The Libyan Meteorological Authority said that entire neighborhoods in Derna disappeared, and their residents were swept away by water after two old dams collapsed, making the situation catastrophic and out of control.

The Libyan government declared 3 days of mourning in all affected cities. He described them as "disaster areas."

According to the NASA Earth Observatory images by Wanmei Liang, using MODIS data from NASA EOSDIS LANCE and GIBS/Worldview, a Mediterranean cyclone contributed to deadly flooding along the Libya's coastline.

Story by Emily Cassidy, http://earthobservatory.nasa.gov/about/Emily-cassidy

Published September 13, 2023 Data acquired September 10, 2023.

Source: Terra > MODIS [5] https://visibleearth.nasa.gov/images/151826/torrential-rain-wreaks-havoc-in-libya/151828w

The collapse of two dams upstream in Libya's eastern city of Derna on Sunday night, the 10th of September, was the cause. The death toll from the catastrophic event has risen to over 8000 people, according to local officials.

DOI: http://dx.doi.org/10.5772/intechopen.1003099

Figure 1.
Animation shows the Derna Valley in Lybia before the flooding Daniel. Source: European Space Agency (ESA). https://www.geospatialworld.net/prime/satellite-imagery-libya-floods/

Figure 2.
Satellite imagery of the Libya flooding Daniel, September 2023. https://visibleearth.nasa.gov/images/151826/torrential-rain-wreaks-havoc-in-libya/151828w

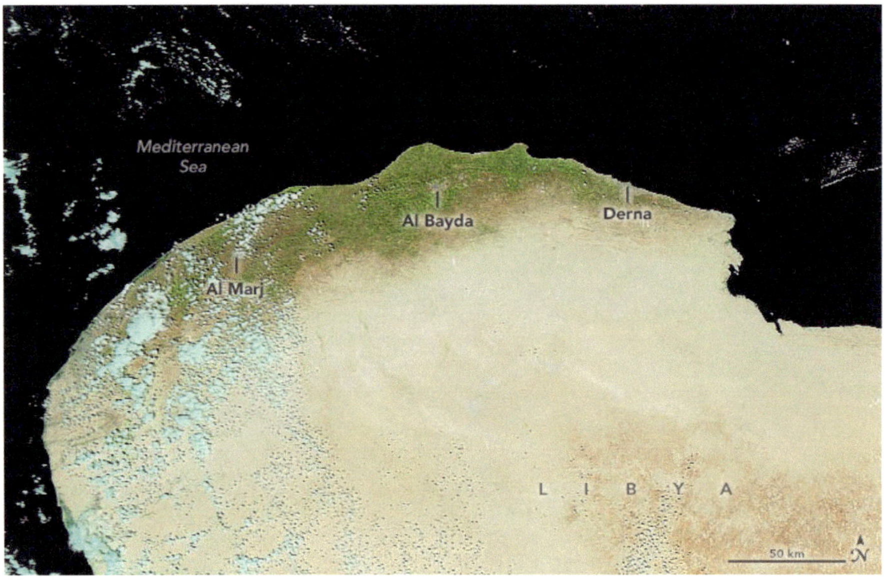

Figure 3.
Libya before flooding Daniel. Data acquired September 7, 2023. https://visibleearth.nasa.gov/images/151826/torrential-rain-wreaks-havoc-in-libya/151828w

Wadi Derna is a river valley in Libya that leads down from the Jebel Akhdar mountains to the port city of Derna. Like many other wadis in North Africa, it is an intermittent riverbed that for much of its length contains water only when heavy rain occurs [6]. It is 75 kilometers (47 mi) long [7] and drains a drainage basin of 575 km^2.

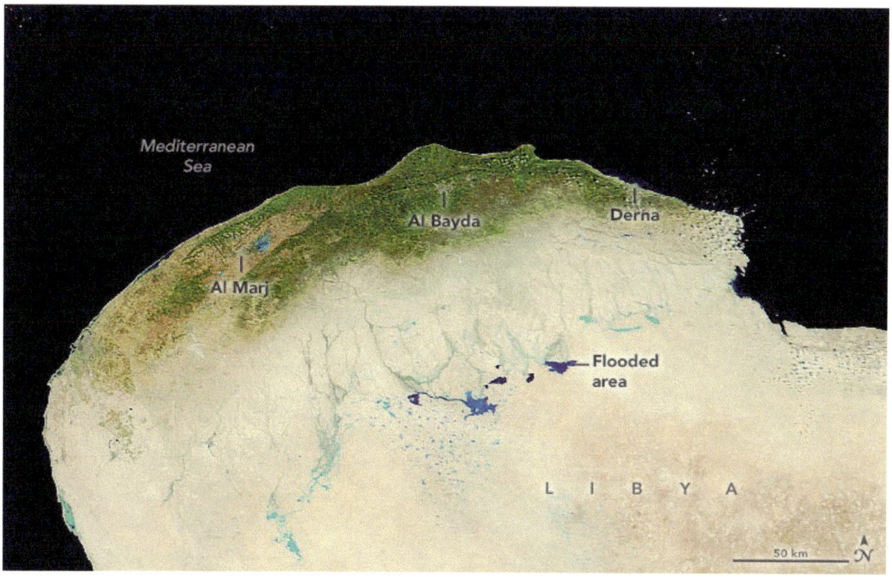

Figure 4.
Libya after flooding Daniel. Data acquired September 13, 2023. https://visibleearth.nasa.gov/images/151826/torrential-rain-wreaks-havoc-in-libya/151828w

Figure 5.
Satellite imagery shows Libya before the flood havoc. Before Sunday night, September 10, 2023. https://www.geospatialworld.net/prime/satellite-imagery-libya-floods/

In September 2023, against the backdrop of the civil war, the rainfall from Storm Daniel led to the collapse of two dams across the river, the Derna dam and the Mansour dam, causing catastrophic flooding in the city of Derna, which killed at least 11,300 people [8–10]. It was the second deadliest dam failure in history.

Figure 6.
Satellite imagery shows flood havoc in Libya. The collapse of two dams upstream in Libya's eastern city of Derna on Sunday night, the 10th of September, was the cause. The death toll from the catastrophic event has risen to over 8000 people, according to local officials. https://www.geospatialworld.net/prime/satellite-imagery-libya-floods/

Figure 7.
Satellite imagery shows flood havoc in Libya. The collapse of two dams upstream in Libya's eastern city of Derna on Sunday night, the 10th of September, was the cause. The death toll from the catastrophic event has risen to over 8000 people, according to local officials. https://www.geospatialworld.net/prime/satellite-imagery-libya-floods

A quarter of Derna city was swept into the sea following Storm Daniel's visit to the northern coast of Libya. The resulting flash flood swept away buildings, people inside them, and vehicles. The floods have also caused widespread damage to infrastructure, including hospitals, schools, and homes.

The World Food Programme has said that the floods could displace up to 100,000 people and could lead to a food crisis in the region (**Figures 1–8**) [11].

Figure 8.
Aerial view of Derna, Libya, on September 10–11, 2023. AYMAN AL-SAHILI/REUTERS.

Figure 9.
A LIDAR (light detection and ranging) image.

9. Advanced remote sensing applications for preservation of Archeological sites

The Earth Observing System (EOS) platform is an advanced cloud platform that has been used to search, analyze, store and visualize geospatial data. This online

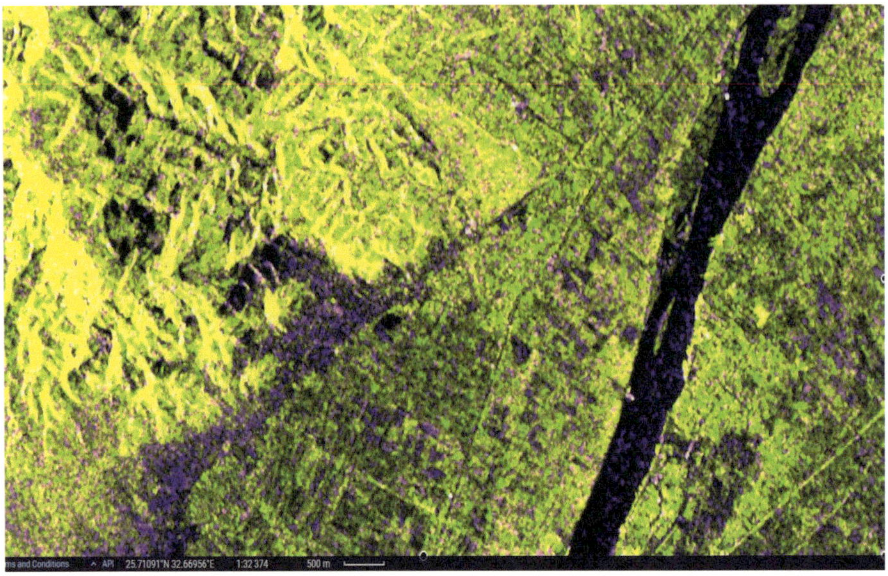

Figure 10.
LIDAR image; the ratio "VV, VH, VV/VH" combination works well for distinguishing water objects, open soils, moistened agricultural fields, vegetation, and urban areas.

Figure 11.
The classification map that includes four different classes for clouds.

digital platform provides powerful tools to search, process, and analyze the large amounts of satellite data generated. As a result, valuable insights allowed us to respond quickly to changes in our area of interest (Valley of the Kings (KV) in Luxor, Egypt).

ELiT (EOS LiDAR Tool) is a cutting-edge product based on complex algorithms for modeling and analysis of the urban environment. I presented the original cross-functional approach to processing LiDAR data in the Valley of the Kings.

Figure 12.
This band combination is good for picking out land from water.

DEM creates a digital elevation model—representing bare earth terrain with regularly spaced z-values. The DEM tool allowed us to build a digital elevation model that represents the bare terrain of the Valley of the Kings (KV). This topographic surface is based exclusively on the Airborne LiDAR Point Cloud and can be obtained in the form of either a DSM (Digital Surface Model) that includes buildings, woody vegetation, and other objects or a DTM (Digital Terrain Model) with all natural and human elements. Objects made above ground are removed. The topographic data used in this study is a digital elevation model where each cell consists of X and Y coordinate information and a Z evaluation (**Figures 9–12**) [12–14].

Author details

Sayed Hemeda
Cairo University, Cairo, Egypt

*Address all correspondence to: hemeda_200x@yahoo.com

References

[1] Toure S, Diop O, Kpalma K, Maiga AS. Shoreline detection using optical remote sensing: A review. ISPRS International Journal of Geo-Information. 2019;**8**(2):75. DOI: 10.3390/ijgi8020075

[2] King MD, Platnick S. The earth observing system (EOS). Comprehensive Remote Sensing [Internet]. Vol. 1. 2018. pp. 7-26. [cited October 01, 2023]. Available from: https://ntrs.nasa.gov/citations/20180003045

[3] Butt MJ, Kelly R. Estimation of snow depth in the UK using the HUT snow emission model. International Journal of Remote Sensing. 2008;**29**(14):4249-4267. DOI: 10.1080/01431160801891754

[4] Hemeda S. Influences of bulk structure of Calcarenitic rocks on water storage and transfer in order to assess durability and climate change impact. Heritage Science. 2023;**11**:118. DOI: 10.1186/s40494-023-00949-w

[5] Torrential Rain Wreaks Havoc in Libya. [Internet]. visibleearth.nasa.gov. 2023. [cited October 01, 2023]. Available from: https://visibleearth.nasa.gov/images/151826/torrential-rain-wreaks-havoc-in-libya/151828w

[6] Mellen R, Karklis L, Granados S, Ledur J, Stillman D. Mapping Why Libya's Floods Were So Deadly. Washington, DC, US: Washington Post; 12 Sep 2023. [Accessed: September 13, 2023]

[7] McBurney CBM, Hey RW. VII, Tufaceous deposits. Prehistory and Pleistocene Geology in Cyrenaican Libya. Washington, DC, US: Cambridge University Press; 1955. Available from: books.google.co.uk. ISBN: 9780521056243 [Accessed: September 14, 2023]

[8] Death Toll Hits 11,300 in Libyan City Destroyed by Floods. NBC News; 2023-09-14. [Accessed: September 15, 2023]

[9] Bodies Wash Ashore in Libya as Devastated City Races to Count its Dead. NBC News; 2023-09-13. [Accessed: September 14, 2023]

[10] Derna W. [Internet]. Wikipedia. 2023. Available from: https://en.wikipedia.org/wiki/Wadi_Derna

[11] Jacob J. In pictures: Satellite imagery shows flood havoc in Libya. Geospatial World. 2023. [Internet] [cited October 01, 2023]. Available from: https://www.geospatialworld.net/prime/satellite-imagery-libya-floods/

[12] Hemeda S. Geo-environmental monitoring and 3D finite elements stability analysis for site investigation of underground monuments. Horemheb tomb (KV57), Luxor, Egypt. Heritage Science. 2021;**9**:17. DOI: 10.1186/s40494-021-00487-3

[13] Hemeda S, Sonbol A. Sustainability problems of the Giza pyramids. Heritage Science. 2020;**8**:8. DOI: 10.1186/s40494-020-0356-9

[14] Hemeda S. Engineering failure analysis and design of support system for ancient Egyptian monuments in valley of the kings, Luxor, Egypt. Geoenvironmental Disasters. 2018;**5**:12. DOI: 10.1186/s40677-018-0100-x

Chapter 2

Mineral Mapping through Advanced Airborne Hyperspectral Remote Sensing Techniques

Nisha Rani, Sumit Kumar Ahirwar, V.M. Anoop and K.V. Krishnamurthy

Abstract

An increased number of spectral channels in the visible to near-infrared region (VNIR) and short-wave infrared region (SWIR) in hyperspectral sensors help in the identification of distinct variations in minerals. Airborne Visible Infrared Imaging Spectrometer-Next Generation (AVIRIS-NG) hyperspectral sensor with 425 spectral bands is useful for mineral Constrained Energy Minimization (CEM), Spectral Angle Mapper (SAM), and Adaptive Coherence Estimator (ACE) advanced tools utilized to detect target minerals in the study area for mineral exploration. Minimum Noise Fraction (MNF), ACE, SAM and CEM advanced hyperspectral remote sensing techniques were applied to the AVIRIS-NG airborne data to map the kaolinite, dolomite, chlorite and hematite. In the MNF images, some areas are prominently highlighted and identified as areas of interest. CEM algorithm has classified the four minerals better than the ACE and SAM. The potential zones identified from the remote sensing data were verified, and abandoned old mines are present in three locations out of four. These locations are picked up well in different MNF images. In the mapped area, carbonate is present in the form of altered minerals in metabasalt, and iron leached surface indicates hematite mineral. The remote sensing outputs well also supported by the petrographic studies. The approach used in this study has great implications for mineral mapping.

Keywords: AVIRIS-NG, alteration, CEM, ACE, minerals, potential zone, gold

1. Introduction

The search for mineral deposits has been a continuous endeavour throughout history. Due to the rapid rate of growth, known resources are depleting at a much faster rate, necessitating the development of novel strategies for the search and exploration of additional resources at an equally rapid rate. Incorporating advanced techniques to identify and delineate new hidden deposits has become increasingly difficult. Remote sensing techniques can fulfill the required demand to locate new potential zones of mineralization within a relatively short time.

With the advancement in remote sensing techniques from multispectral to hyperspectral, remote sensing has played a key role in the field of geosciences, especially in mineral exploration [1]. An increased number of spectral channels in the visible to near-infrared region (VNIR) and short-wave infrared region (SWIR) in hyperspectral sensors helps in the identification of distinct variations in minerals [2]. Alteration minerals associated with certain types of potential mineralized zones usually show their distinct features in the VNIR and SWIR regions (0.4–2.5 μm) of the electromagnetic spectrum [3]. As a result of the exclusive amalgamation of spatially contiguous spectra and spectrally contiguous images, hyperspectral remote sensors are able to map alteration minerals spatially. The spectral range 0.4–2.5 μm provides abundant information about a wide variety of Earth-surface minerals, and the 2.0–2.5 μm range covers spectral characteristics of hydroxyl-bearing minerals, sulphates, and carbonates that are found in many geologic units and hydrothermal systems [4]. For quantitative and qualitative analysis and mapping, hyperspectral sensors with coherent band parameters are vital [2].

Hyperspectral remote sensing techniques have been used to gather mineral information for over three decades, and its application in the Indian sub-continent is still at a nascent stage. There are several reasons why the hyperspectral technique is not widely used, and the main reason is lack of spatial and aerial coverage and another non-availability of the data. Airborne Visible Infrared Imaging Spectrometer-Next Generation (AVIRIS-NG) hyperspectral data was acquired over the Indian sub-continent in two phases by a joint mission of the Indian Space Research Organization (ISRO) and National Aeronautics and Space Administration (NASA) from 2015 to 2016 and 2018 to 2019. In the AVIRIS-NG data, there are about 430 narrow continuous spectral bands in VNIR and SWIR from 0.38 to 2.5 μm range at 5 nm intervals, and pixel resolution varies from 4 to 8 m for a flight altitude of 4–8 km for a swath of 4–6 km. AVIRIS-NG is a recent trend in airborne hyperspectral remote sensing that can be considered as the most advanced data for mineral identification and mapping [1]. In the present study, AVIRIS-NG acquired over the west of Gadag Schist Belt (GSB), Southern India, is analyzed for mineral mapping using advanced Minimum Noise Fraction (MNF), Constrained Energy Minimization (CEM), Spectral Angle Mapper (SAM) and Adaptive Coherence Estimator (ACE) techniques.

2. Study area and geological setting

The study area is located west of Gadag Schist belt (GSB), Karnataka, India (150 12′ 0″ to 150 20′ 0" N and 750 34′ 0″ to 750 40′ 0″ E; **Figure 1a**). The study area is the geologically northern part of the Chitradurga Schist Belt (CSB) of Western Dharwar Craton (WDC) of Southern India [3]. The schist belt is mostly occupied by metabasalts, metavolcanics, granite gneiss, conglomerate, greywacke, argillic, and banded iron formation rocks [5]. The belt trends NW-SE in strike length of around 50 km with a maximum width of around 22 km is between Shirahatti in the west and Dambal in the east and is a site of ancient mining activity and old mines are located 15–25 km due south of the Gadag city [6]. Gold mineralization is associated with several prominent shears, striking NNW-SSE and is controlled by shears and marked by wall rock alterations [5]. Geological Survey of India (GSI) systematically explored the gold-bearing tracts in the western half of the GSB [6]. The study area is mainly occupied by the variants of metabasalt,

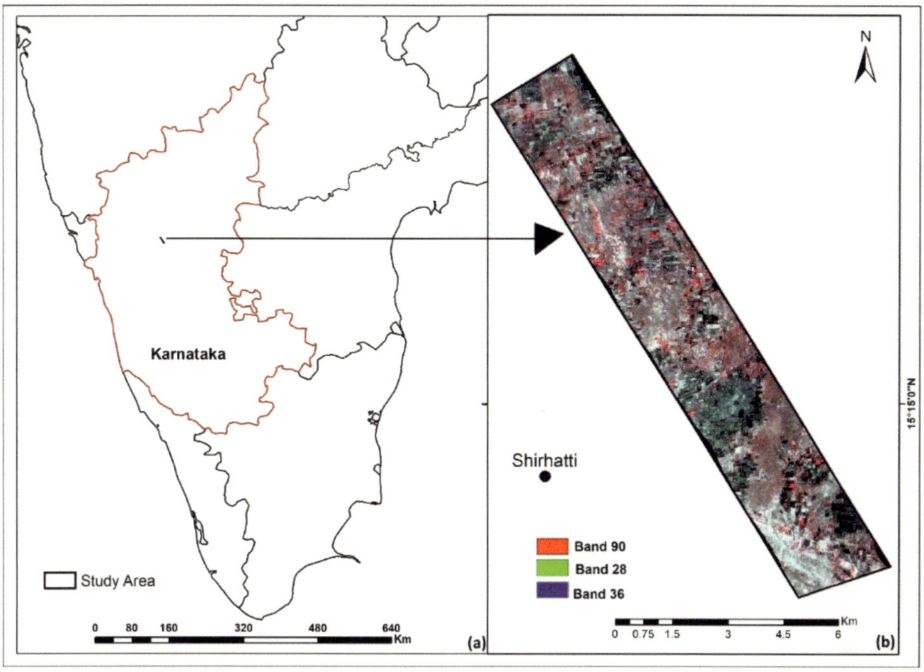

Figure 1.
Uninterpreted false color composite (FCC) (red band-90, green band-28 and blue band-36) of AVIRIS-NG airborne data of the study area (black box). The study area is part of Karnataka state (red boundary line) of India.

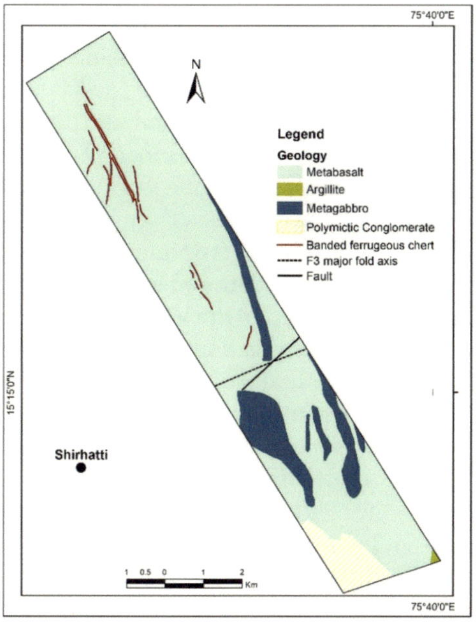

Figure 2.
Geological map of the study area (modified after the geological survey of India). The major part of the area is occupied by metabasalt, and in the central and southern part, metagabbro. Banded ferruginous chert, occupying the northern part trending NW-SE. (the map is modified from the URL: http://gsi.gov.in/bhukosh.)

metagabbro, polymictic conglomerate, argillite and banded ferrugenous chert and the of chert band is NW-SE. South central part is having traces of F3 major fold axis and NE-SW trending fault. Source **Figure 2**.

3. Methodology

For the present study, AVIRIS-NG (scene id: ang20180324t054258_rfl_v2s1) was downloaded from https://vedas.sac.gov.in. The AVIRIS-NG is level 2 reflectance data, acquired in 2A–2B phase during 2018–2019. Before applying advanced hyperspectral remote sensing techniques on the data, the bands were resized by removing bands which do not contain any information from the raw data. Total 328 VNIR and SWIR channels out of 425 were considered for the analysis. To map hydroxyl, carbonate and iron-bearing minerals, techniques like Minimum Noise Fraction (MNF) and spectral classifications like Constrained Energy Minimization (CEM), Spectral Angle Mapper (SAM), and Adaptive Coherence Estimator (ACE) were adopted. **Figure 3** shows the methodology followed in the present study.

3.1 Minimum noise fraction (MNF)

Sometimes, unprocessed imageries do not produce favoured outcomes (**Figure 4a** and **b**); therefore, to extract better and desired outcomes application of advanced enhancement techniques is required. MNF is known to be an effective tool for reducing the dimensionality of large datasets to a small number of components that retain most of the information. MNF is useful for discerning between noise and signals in a dataset [7]. This transformation is similar to Principal Component Analysis (PCA) and is a very powerful tool in lithological, structural and alteration zone mapping. MNF is used to determine how much noise is in the image data, separate noise from the image, and reduce the computation required for further processing [8]. MNF includes two steps: the first transformation is called noise-whitening and is based on an estimated noise covariance matrix that aims at decorrelation and rescaling the noise in the data. This transformation is capable of determining the inherent dimensionality of an image [7]. The second step derives the principal components from the noise-free data. Then the data can be split into two parts. One part is associated with large eigenvalues, the other part has eigenvalues close to 1, and the image is dominated by noise. Using data with large eigenvalues separates noise from the data and improves spectral results [9].

First, the MNF technique was applied to the VNIR & SWIR bands of AVIRIS-NG data with 30 output bands for the identification of different lithology and alteration zones in the study area. MNF image obtained by applying MNF transformation on AVIRIS-NG-VNIR-SWIR bands (**Figures 5** and **6**). In the False Color Composite (FCC) image of MNF bands (R-7: G-4: B-10), greenish color dominate in the north, shades of magenta in visible in the central and southern part, the prominent hue of magenta-blue are confined in the south-west part (**Figure 5a**). In the FCC image of MNF bands (R-12: G-10: B-9), light pink prevails in the north, yellow is the south part, and green is distributed in the entire area (**Figure 5b**). In the FCC image of MNF bands (R-6: G-5: B-4), prominent green color in the southern part of the study area marks the boundary in two different litho-units, highlighting linear and curvilinear features **Figure 6a**. In the FCC image of MNF bands (R-4: G-8: B-5), deep magenta color features are prominent in the northern part and light blue in the south (**Figure 6b**). In the FCC image (**Figure 7**) of MNF bands (R-6: G-7: B-8), yellow to rust color dominates in

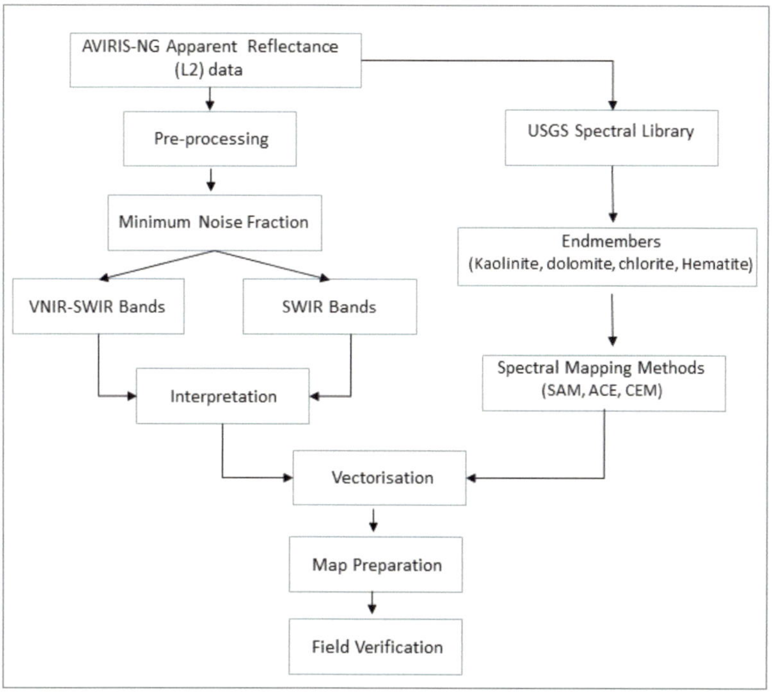

Figure 3.
Methodology followed in the present study to map the minerals using advanced hyperspectral techniques.

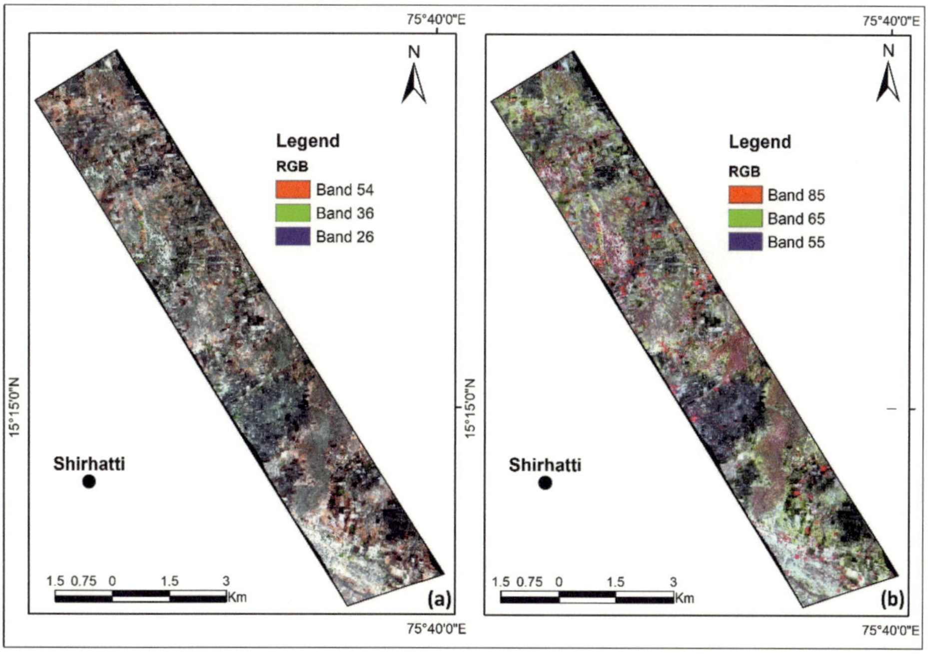

Figure 4.
Unprocessed AVIRIS-NG imagery of the study area. (a) Is true colour image (R-54: G-36: B-26); and (b) is colour infrared image (R-85: G-65: B-55).

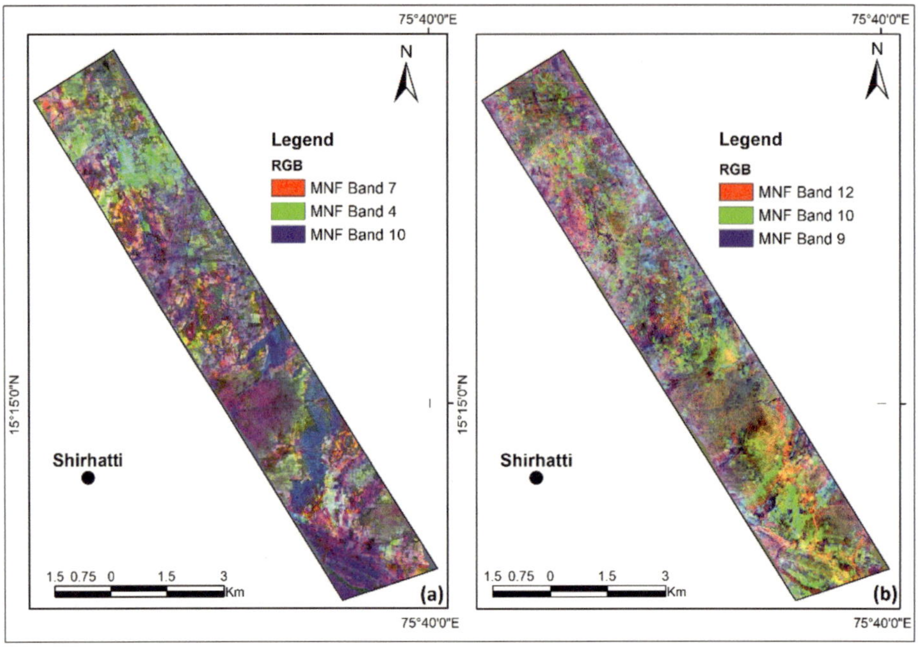

Figure 5.
MNF image obtained by applying MNF transformation on AVIRIS-NG-VNIR-SWIR bands. (a) FCC image of MNF bands (R-7: G-4: B-10 and visual interpretation of this figure shows greenish color in north, shades of magenta in the central and southern parts and a hue of magenta-blue confined in the south-west part. (b) In the FCC image of MNF bands (R-12: G-10: B-9), light pink prevails on the north side and yellow on the south part, whereas green is distributed in the entire area.

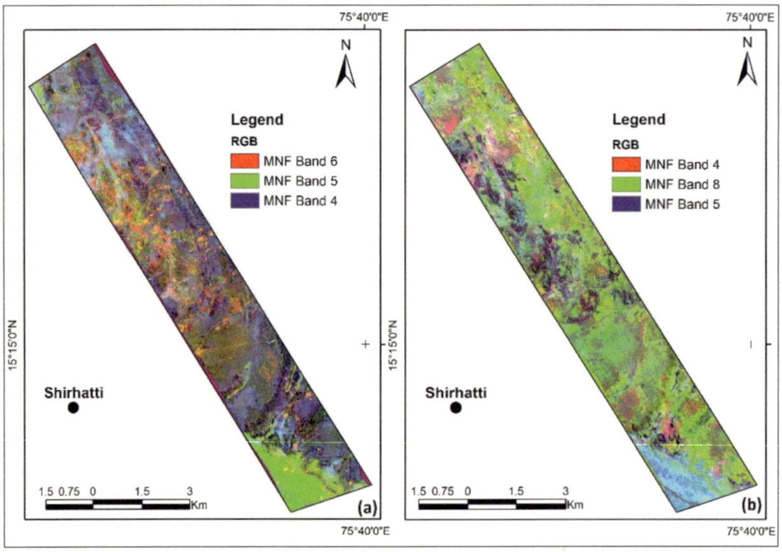

Figure 6.
MNF image obtained by applying MNF transformation on AVIRIS-NG-VNIR-SWIR bands. (a) In the FCC image of MNF bands (R-6: G-5: B-4), in the southernmost corner, a distinct boundary in green and blue color is prominent, indicating two different litho-unit. (b) In the FCC image of MNF bands (R-4: G-8: B-5), deep magenta color is prominent and confined in the northern-western part and cyan color in south.

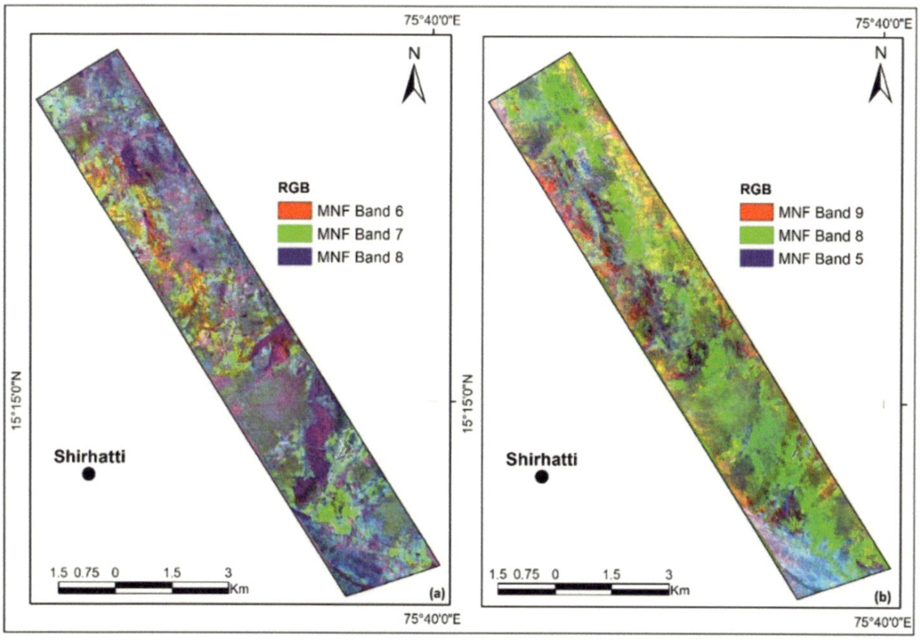

Figure 7.
MNF image obtained by applying MNF transformation on AVIRIS-NG-VNIR-SWIR bands. (a) In the FCC image of MNF bands (R-6: G-7: B-8), yellow to rust color, in a linear pattern, dominates in the western and central parts. In the eastern part of the study area, magenta is prominent along with green, whereas blue hue is in the south. (b) FCC image of MNF bands (R-9: G-8: B-5), maroon is prevailing in the northern-western part, trending in NW-SE direction, the eastern margin is in yellow whereas green is common in the entire area. The south-west corner is highlighted in cyan and pink color.

a linear pattern in the western and central parts. In the eastern part of the study area, magenta is prominent along with green, whereas blue hue is in the south (**Figure 7a**). In the FCC image of MNF bands (R-9: G-8: B-5), maroon is prevailing in the western part in NW-SE direction, yellow in the eastern margin whereas green is common in entire area. South-west corner is highlighted in light blue to pink color (**Figure 7b**).

Spectral range from 2.0 to 2.5 μm, provides abundant information about hydroxyl-bearing minerals, sulfates and carbonates that are found in many geologic units and hydrothermal systems [4], SWIR bands of AVIRIS-NG were considered for mapping different lithology which may be associated with some potential mineralized zone (**Figure 8**). In the **Figure 8a**, FCC image of MNF bands (R-6: G-5: B-7); a very prominent trend in NW-SE direction is highlighted in the northern part. In the southern tip of the study area, a NW-SE prominent boundary between peach and green color is visible, indicating contact of two different lithologies. FCC image of MNF bands (R-7: G-5: B-3) also indicates the prominent trend of NW-SE magenta color along with yellow and yellow colours in the north (**Figure 8b**). A few spots in magenta are visible in the south part, whereas lithological contact in the southern part is not prominent as in the MNF image R-6: G-5: B-7.

3.2 Classification techniques

Supervised classification is used to cluster pixels in the dataset into classes corresponding to user-defined training classes. Training classes are groups of pixels

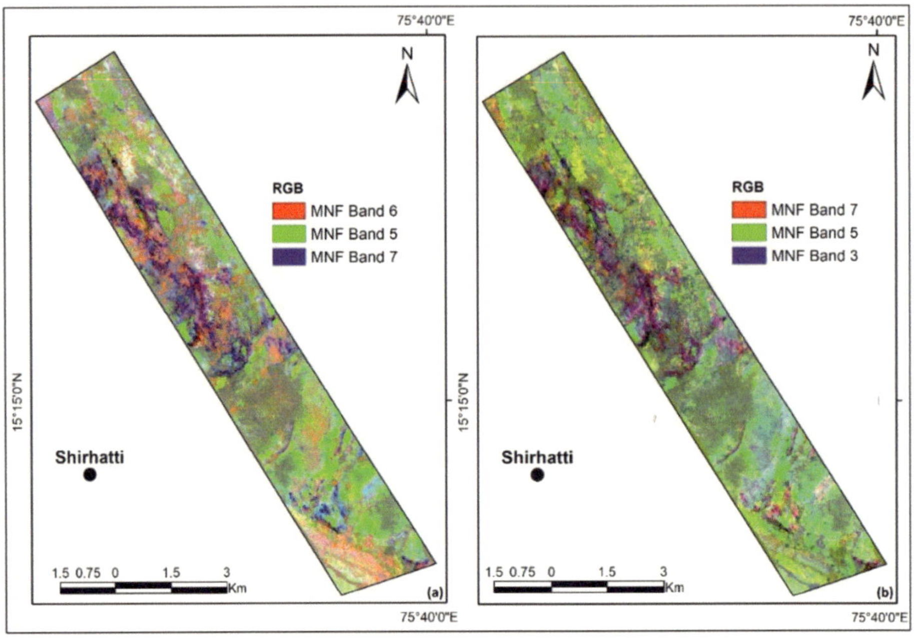

Figure 8.
MNF image obtained by applying MNF transformation on AVIRIS-NG-SWIR bands. (a) In the FCC image of MNF bands (R-6: G-5: B-7), a very prominent trend in the NW-SE direction is highlighted in the north-western part. An NW-SE prominent boundary between peach and green color is visible in the southern tip of the study area, indicating contact between two different lithologies. (b) FCC image of MNF bands (R-7: G-5: B-3) also indicates a prominent trend in NW-SE of magenta and yellow in the north-western part. In the south, light magenta is visible at a few locations, whereas lithological contact in the southern part is not highlighted as in other MNF images.

(Region of Interests-ROIs) or individual spectra of minerals. There is various classification method to map the mineral using hyperspectral remote sensing data. Advanced spectral classification methods such as Constrained Energy Minimization (CEM), Spectral Angle Mapper (SAM), and Adaptive Coherence Estimator (ACE) techniques are powerful tools for detecting target minerals in potential zones for mineral exploration.

3.2.1 Constrained energy minimization (CEM)

CEM is an extension to spectral unmixing, established by Farrand and Harsanyi [10]. This technique has been widely used in various applications, such as geological mapping, agriculture management and medical image processing for target detection. CEM is originally derived from a linearly constrained minimum variance adaptive beam-forming in the field of signal processing. This technique keeps output energy constant at the same time suppressing the background to a minimum level [11, 12]. CEM is widely utilized for target detection in hyperspectral imagery. It detects the desired target signal source using a unity constraint while suppressing noise and unknown signal sources by minimizing the average output power [13]. Using partial unmixing based on the estimate of sample correlation matrix highlights target abundances [14]. In order to suppress undesired background and improve the target signature, the total output energy of all the pixels in the imagery is minimized, and the total energy of each individual pixel is projected to be equal to one. The resultant

CEM image is the vector component, containing abundance information [15]. CEM algorithm requires the signature of the target only, and prior knowledge is not obligatory. An additional benefit of CEM is that as various signal sources cannot be recognized or observed with the bare eye, several materials may be detected by sensors leading to false [11, 12].

3.2.2 Spectral angle mapper (SAM)

SAM classification is based on a hypothesis that in satellite imagery, a single pixel represents one particular ground material and can be exclusively assigned to only one ground cover class [16]. The SAM classification technique classifies the pixels based on similar spectral properties with reference spectra. It utilizes endmember spectra that could be extracted directly from the image or spectral library [17]. The algorithm determines the spectral similarity between two spectra by calculating the angle among the spectra and treating them as vectors in a space with dimensionality equal to the number of bands [18]. These vectors are generated by connection of reflectance spectrum point with the origin. SAM is based on the spectral shape outline, whereas conventional classifiers are based on the statistical distribution of the pattern [15]. Smaller angles represent closer matches to the reference spectrum. Pixels further away than the specified maximum angle threshold in radians are not classified.

3.2.3 Adaptive coherence estimator (ACE)

The ACE method determines if a pixel spectrum certainly comprises of a known target signature. It gives the best results when the background conditions are variable and unknown. ACE algorithm is a uniformly most powerful invariant detection statistic that is derived from the Generalized Likelihood Ratio (GLR) approach [19]. The ACE is invariant to the relative scaling of input spectra and has a Constant False Alarm Rate (CFAR) with respect to such scaling. Similar to CEM and SAM, ACE does not require knowledge of all the end members within an image scene. ACE classifier considers both the statistical model as well as subspace projection model into account. Calculating the cosine square of the angle between two spectra yields information about how similar the test and reference spectra are related to each other. Since this approach is noise-sensitive, it is necessary to eliminate noise before target detection [20].

4. Data processing

In this investigation, the reference spectra of selected endmember minerals, for implementing CEM, SAM and ACE algorithms were extracted from the USGS spectral library. These reference spectra were convolved to the spectral response function of AVIRIS-NG's VNIR and SWIR bands. Selected endmembers spectra in this study are kaolinite, dolomite, chlorite and hematite (**Figure 9a–c**).

The CEM, SAM and ACE algorithms were implemented on the VNIR-SWIR bands of AVIRIS-NG data separately (**Figures 10** and **11**) to obtain the desired outcome and results appear as a series of grayscale images for each selected end member. These algorithms were applied to the selected spatial subset AVIRIS-NG data of the study area to map the subpixel distribution of minerals collected from the USGS library. The RGB color composite was analyzed and found non-informative. The resultant gray image of each mineral was classified, and the grayscale in images shows high DN

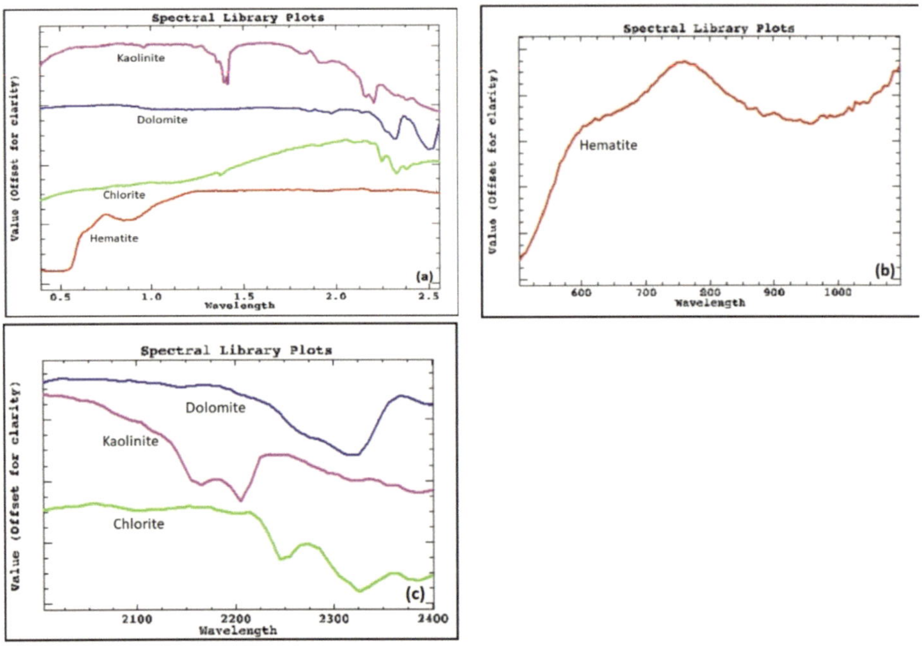

Figure 9.
Convolved USGS spectral library profiles with AVIRIS-NG Spectral response function using ENVI, (a) refers to the USGS spectra for the alteration minerals (kaolinite, dolomite and chlorite and hematite), (b) shows the spectral curves of hematite after resampled with VNIR bands of AVIRIS-NG sensor spectral response function, (c) represents the spectral curve of dolomite, kaolinite, and chlorite after resampled with SWIR bands of AVIRIS-NG sensor spectral response function.

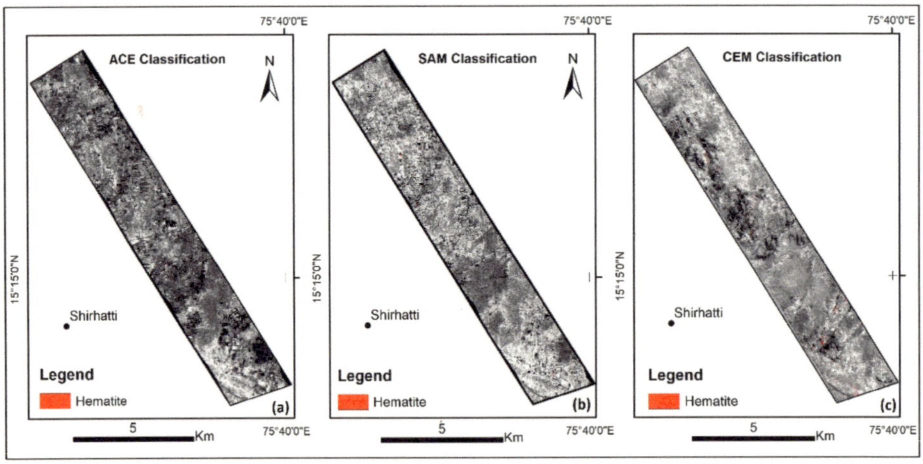

Figure 10.
The spatial distribution of hematite mineral in gray image, obtained by applying ACE, SAM and CEM spectral classification techniques on VNIR bands of AVIRIS-NG data of the study area. (a) Derived from ACE, (b) resulted by SAM and (c) through the CEM algorithm.

(digital number) values of the sub-pixel abundance of the mineral in the image. To ascertain the target mineral, each gray scale image was classified in gray scale through a density slicing approach. The highest value of density slicing was considered,

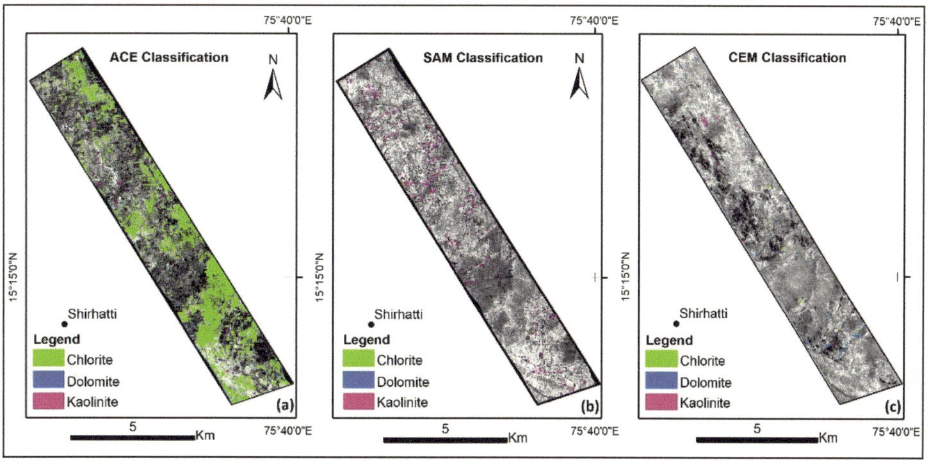

Figure 11.
The spatial distribution of kaolinite, dolomite and chlorite minerals in gray image, obtained by applying ACE, SAM and CEM spectral classification techniques on SWIR bands of AVIRIS-NG data of the study area. (a) Derived from ACE, (b) resulted from SAM and (c) through the CEM algorithm.

and each value in the resultant image represents the sub-pixel abundance of the target alteration mineral in each pixel. Subtle differences between altered and non-altered rocks and hydrothermal alteration minerals were detected and mapped in the study area.

Figure 10a–c shows the spatial distribution of hematite mineral on gray image of AVIRIS-NG data obtained by applying ACE, SAM and CEM spectral classification techniques on VNIR bands of AVIRIS-NG data of the study area. **Figure 11a–c** displays the spatial distribution of kaolinite, dolomite and chlorite minerals on gray image of AVIRIS-NG data, derived by applying ACE, SAM and CEM spectral algorithms on SWIR bands of AVIRIS-NG. **Figure 12** demonstrate combined

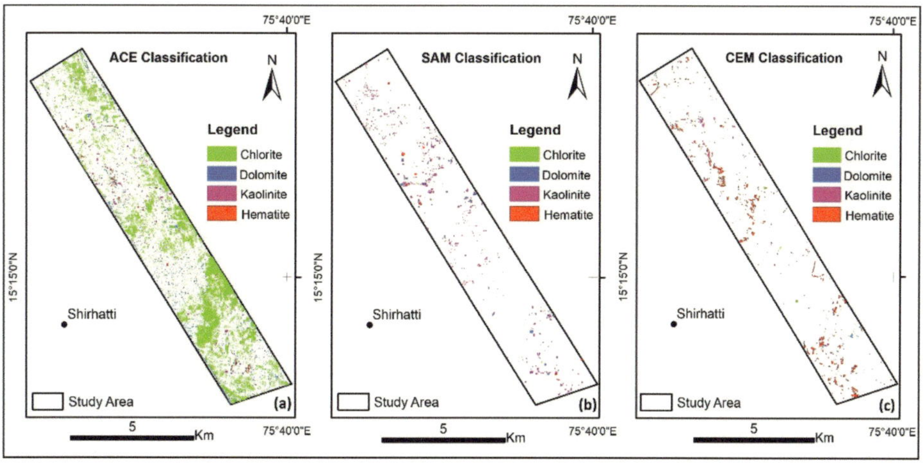

Figure 12.
Combined resultant mineral maps derived from ACE and SAM algorithm applied on VNIR-SWIR bands of AVIRIS-NG image of the study area to map the minerals. (a–c) The spatial distribution of kaolinite, dolomite, chlorite and hematite minerals obtained by ACE, SAM and CEM spectral classification methods.

resultant mineral maps derived from ACE and SAM algorithm applied on VNIR and SWIR bands of AVIRIS-NG image of the study area to map the minerals separately. **Figure 12a–c** shows the spatial distribution of kaolinite, dolomite, chlorite and hematite minerals obtained by ACE, SAM and CEM spectral classification methods.

ACE and SAM did not perform well despite analyzing different threshold values. The SAM rule image was used to produce a classified image map. But, the image map contains a very noisy background, and classified mineral zones was not discernible from the background (**Figure 12**). **Figure 12a** shows the resultant image map derived from ACE algorithm in the study area. ACE has classified most of the study area as chlorite mineral. Hematite shows the concentration in the northern part, and the roads present in the area are also identified as target mineral hematite. Hematite and chlorite could not be recognized by the SAM algorithm as expected (**Figure 12b**). CEM was applied to map the subpixel distribution of kaolinite, dolomite, chlorite and hematite in the study area. The resultant gray image of each mineral was classified, and the grayscale in images shows high values of the sub-pixel abundance of the mineral in the image. Comparative analysis of the spectral classification results (**Figure 12a–c**) shows that all four minerals were well detected and mapped by CEM than ACE and SAM algorithms. The CEM algorithm can detect the detailed surface distribution of alteration minerals in the area better than other algorithms.

For better interpretation, visualization and understanding, results of mapped minerals displayed on the gray band of AVIRIS-NG and outline of the geology (**Figure 13**). **Figure 13a** shows the spatial distribution of the minerals on gray image

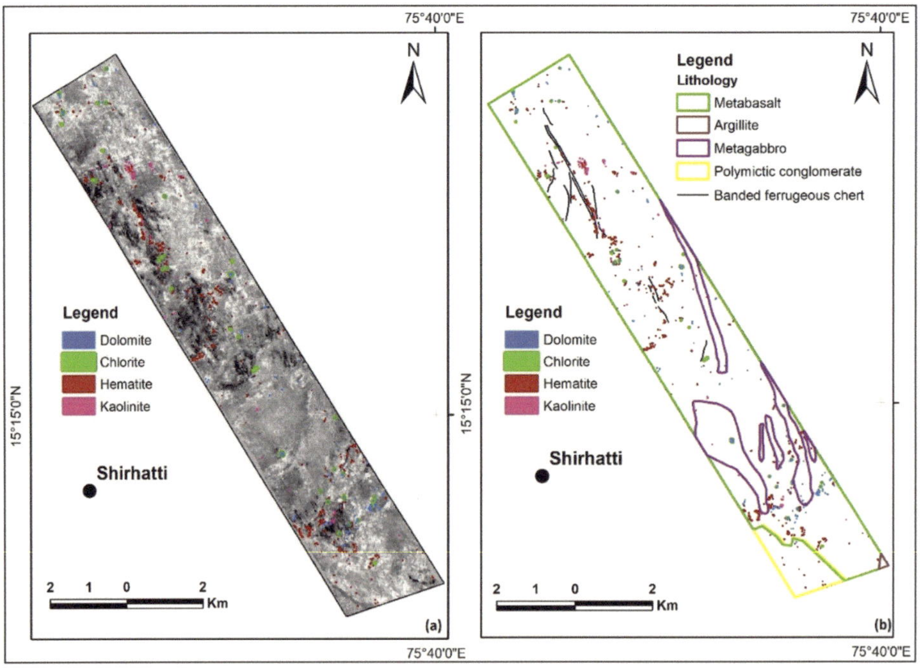

Figure 13.
The spatial distribution of the combined minerals (obtained separately from VNIR and SWIR bands) on gray image and correlation of mapped mineral with the existing geology of the study area. (a) Show the spatial distribution of kaolinite, dolomite, chlorite and hematite minerals on gray image. (b) Shows the correlation of the detected minerals with the known geology of the study area, mapped by GSI.

of AVIRIS-NG data. Visual interpretation indicates that the majority of the mapped minerals are concentrated in the northern part and in some isolated areas in the south part of the research area. **Figure 13b** shows an outline of the geology area mapped by GSI and the results of the mapped minerals. **Figure 13b** attempts to establish the correlation of the detected minerals with known geology. Hematite mineral shows concentration along NW-SE trending banded ferruginous chert hosted in metabasalt. Kaolinite and hematite minerals have an overlapping relationship observed at places. Chlorite, dolomite and some hematite minerals are confined in metagabbro.

The prominent features and information highlighted in the MNF images of AVIRIS-NG were identified as areas of interest and vectorised. The area is well known for gold mineralization, and a database in the GIS platform on gold mineralization was prepared. The mapped minerals: kaolinite, dolomite, chlorite, hematite, known locations of gold mineralization and area of interest identified from the MNF image were integrated together to verify the remote sensing outputs in the study area and to identify the potential zones for mineral investigation (**Figure 14**). **Figure 14a** shows an outline of the area of interest (vector) demarcated from different MNF images, old workings and gold locations reported by GSI in the study area superimposed on the MNF image (R-7: G-5: B-3). **Figure 14b** shows an outline of the area of interest (vector) demarcated from different MNF images, old workings & gold locations reported by GSI and the mapped alteration minerals for better visualization and demarcation of the potential areas for field investigation. Old workings and other reported gold mineralization in the area show a good correlation with the mapped minerals through CEM, SAM and ACE classification.

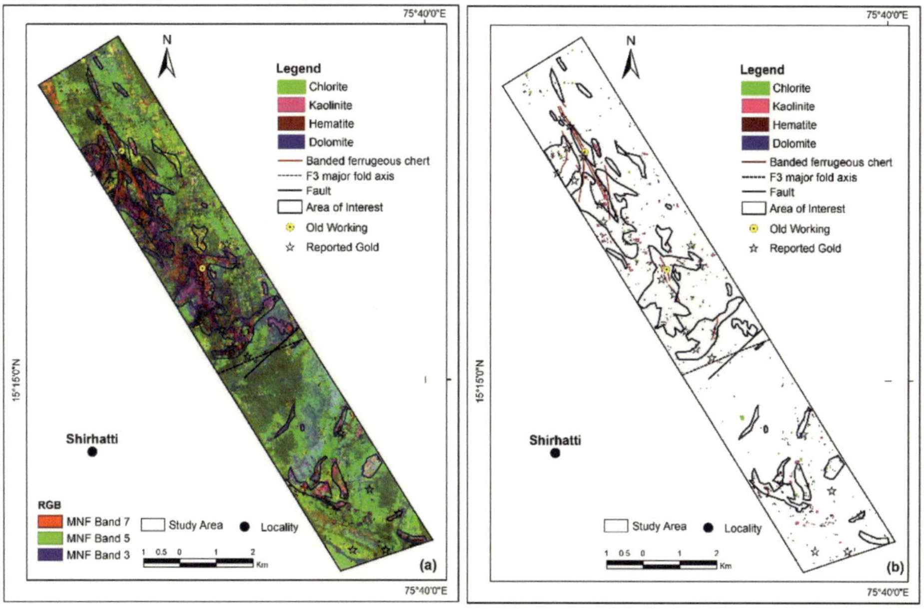

Figure 14.
Shows integration of mapped minerals (kaolinite, dolomite, chlorite, hematite) derived from ACE, SAM & CEM algorithms, area of interest (potential zones) demarcated from different MNF images, the known location of gold mineralization and structures in the area. (a) Show boundaries of potential zones demarcated from different MNF images interpreted from different MNF images along with the known gold mineralization and structures in the area on MNF image (R-7: G-5: B-3) and (b) show boundaries of potential zones demarcated from different MNF images interpreted from different MNF images along with the known gold mineralization and structures in the area for better interpretation, visualization and identification of potential areas for detailed field investigation.

5. Results

Comprehensive fieldwork and petrographic study were conducted in the study area to verify the remote sensing outputs. Based on the integration of all the inputs, the study area was grouped into four: L1, L2, L3 and L4 (marked as black dotted circles, **Figure 15**) that have been surveyed in the field for the verification of remote sensing outputs. Detailed criteria for the selection of each potential zone for field investigation are given in **Table 1**. **Figure 15a** shows four potential areas along with field location on the MNF image (R-9: G-8: B-5), and **Figure 15b** shows four potential areas along with field location for better visualization.

During the field survey, variants of metabasalt from fine to medium-grained, deformed, foliated and pillowed in nature were recorded. Iron leaching on the surface of the rocks is also very common in those areas where BIF is present. In addition, intense carbonate alteration was observed in the metabasalt. In the investigated area, several abandoned old workings were noticed. Pounding marks on the surface of the rocks in the survey area indicate gold mining activities in ancient times. (**Figure 16a–h**). To ascertain the results of the remote sensing techniques, a lab study of the samples collected from the investigated areas was conducted. The results from the remote sensing technique are well supported by the petrographic study (**Figure 17a–c**). Detailed field observations in four potential zones (L1–L4) and petrographic are given in **Table 2**.

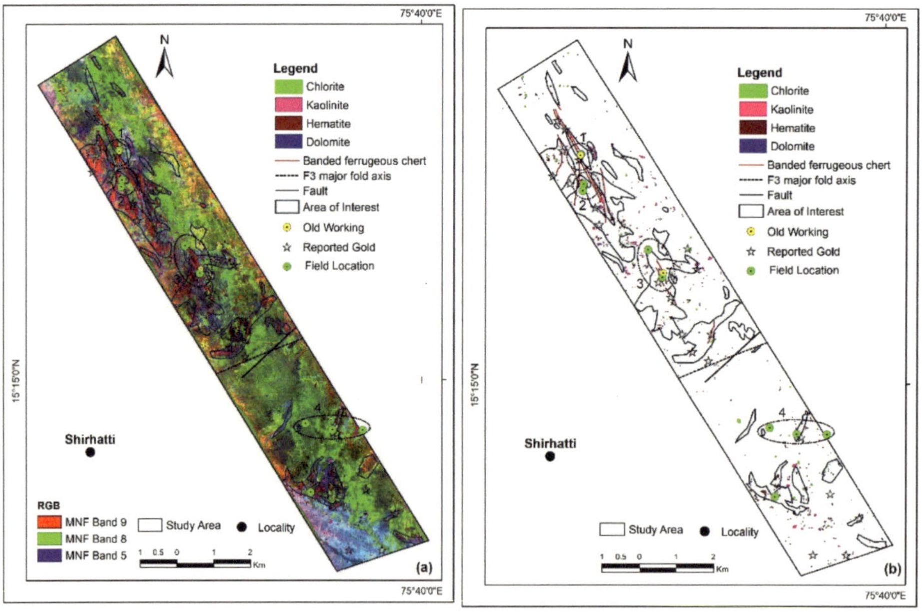

Figure 15.
Four identified areas grouped into four: L1, L2, L3 and L4 for field investigation, marked in black dotted circles along with sample collection locations. (a) Shows four potential areas along with field location on the MNF image (R-9: G-8: B-5), and (b), shows four potential areas along with field location shows without superimposing on image for better visualization.

Potential zones	Correlation with interpreted results	Correlation with geology
L1	Potential zone highlighted in different colors in MNF images and hematite alteration	Banded ferruginous chert, metabasalt and old working
L2	Potential zone highlighted in different colors in MNF images and hematite alteration	Banded ferruginous chert, metabasalt, the occurrence of gold
L3	Potential zone highlighted in different colors in MNF images, hematite and dolomite alteration	Banded ferruginous chert, metabasalt, the occurrence of gold, old working
L4	Potential zone highlighted in different colors in MNF images, hematite, kaolinite, chlorite and dolomite alteration	Metabasalt, the occurrence of gold

Table 1.
Detailed criteria for selection of potential zones for field investigation.

Figure 16.
The area surveyed (L1, L2, L3 and L4) in the field for the verification of remote sensing outputs. In the field, various types of metabasalt are occurring as main rock, minerals altered to carbonate and the surface of rock exposures are iron leached. (a) (L1): Abandoned old working in the area; (b) (L1): Very fine-grained, pillowed metabasalt. (c) (L1): Pounding marks on rock surface indicates gold mining activity in ancient times. (d) (L2): Abandoned site of old working. (e) (L3): Abandoned site of old working. (f) (L3): Metabasalt with a limonitic surface containing grains of pyrite. (g) (L3): Deformed metabasalt. (h) (L4): Very fine-grained, pillowed metabasalt, primary minerals altered to carbonates.

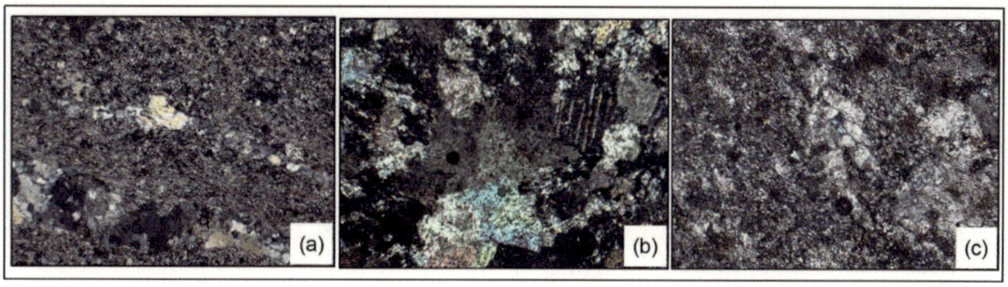

Figure 17.
Different types of alteration minerals in the study area. (a) (L1): Microphotographs show minerals are very fine-grained and composed of carbonate and quartz. Primary minerals are replaced by carbonate. Sericite is also present as a minor mineral. (b) (L3): Microphotograph shows minerals are fine-grained and altered. Calcite is the main component, replacing all the primary minerals indicated by remnants of plagioclase. Sericite and iron oxide are also present. (c) (L4): Microphotograph shows highly altered rock and primary minerals completely replaced by carbonate.

Potential zones	Field observation and petrographic study
L1	During field inspection very fine grained pillowed metabasalt and banded ferruginous chert is observed. Abandoned Hosur Gold Mine is located in this zone and presence of pounding marks on rock surface indicates gold mining activity in ancient times in the area. Hematite alteration is mainly contributed is because of the iron leaching on the and banded ferruginous. In petrographic study, carbonate and sericite alteration minerals recorded
L2	Alteration in metabaslt recorded and abandoned Shirunj Mine is located in this zone. Hematite alteration is due to the presence of iron leached dump
L3	Pillowed, very fine grained metabasalt is altering in carbonate, limonitic rock surface and sulphide recorded. Yellishirur-Venkatapura abandoned mine is present in this zone. Petrographic study shows alteration of primary minerals in calcite and sericite. Iron oxide is also present
L4	Very fine grained, pillowed metabasalt altered in carbonate and medium grained dyke is noticed. Petrographic study shows alteration of primary minerals in carbonate

Table 2.
Location-wise field observations in four identified potential zones and petrographic study.

6. Conclusions

MNF, ACE, SAM and CEM advanced hyperspectral remote sensing techniques were applied on the AVIRIS-NG airborne data to map the kaolinite, dolomite, chlorite, hematite. In the MNF images, some areas prominently highlighted and identified as areas of interest. A comparative study of the results of all three algorithms reveals that CEM algorithm has classified the four minerals better than the ACE and SAM. The potential zones identified from the remote sensing data and the petrographic study were verified in the field. In the field, abandoned mines were present in three of the four potential zones that were well captured in MNF images. Altered minerals identified by CEM, SAM and ACE are well supported by field evidence and petrographic studies. Calcite and hematite alterations recorded in the field which are corresponding to metabasalt-banded ferruginous chert and leached surface. Lab study shows that metabasalt is altering in calcite, and iron oxides are also present. The advanced hyperspectral remote sensing approach used in this study has great implications for mineral mapping.

Acknowledgements

The authors place, on record, their sincere thanks to the Director General, Geological Survey of India, Kolkata and Dy. Director General & Head Mission-V, Geological Survey of India Training Institute, Hyderabad, for permission to publish the work.

Author details

Nisha Rani[1]*, Sumit Kumar Ahirwar[2], V.M. Anoop[1] and K.V. Krishnamurthy[3]

1 Geological Survey of India, Hyderabad, India

2 Geological Survey of India, Bhopal, India

3 Geological Survey of India, Bangalore, India

*Address all correspondence to: nishargsi@gmail.com

References

[1] Tripathi MK, Govil H. Evaluation of AVIRIS-NG hyperspectral images for mineral identification and mapping. Heliyon. 2019;**5**:e02931. DOI: 10.1016/j.heliyon.2019.e02931

[2] Mishra S, Chattoraj SL, Benny A, Sharma Richa U, Ray PKC. AVIRIS-NG data for geological applications in Southeastern Parts of Aravalli Fold Belt, Rajasthan. Proceedings. 2019;**24**:16. DOI: 10.3390/IECG2019-06212

[3] Nisha R, Tejpal S, Mandla VR. Mapping hydrothermal alteration zone through aster data in Gadag Schist Belt of Western Dharwar craton of Karnataka, India. Environmental Earth Sciences. 2020;**79**:526. DOI: 10.1007/s12665-020-09269-9

[4] Kruse FA. Comparison of AVIRIS and hyperion for hyperspectral mineral mapping. In: Presented at the 11th JPL Airborne Geoscience Workshop; 4-8 March 2002; Pasadena. Boulder, Colorado, USA: California 1 Analytical Imaging and Geophysics LLC; 2002

[5] Nisha R, Mandla VR, Tejpal S. Structural and alteration mapping using ASTER imagery and DEM for gold mineralization in the Gadag Schist Belt of Karnataka, India. In: Misra AA, Mukherjee S, editors. Atlas of Structural Geological and Geomorphological Interpretation of Remote Sensing Images. Chichester, West Sussex, UK: John Wiley & Sons Ltd; 2022

[6] Nisha R, Mandla VR, Tejpal S. Spatial distribution of altered minerals in the Gadag Schist Belt (GSB) of Karnataka, southern India using hyperspectral remote sensing data. Geocarto International. 2016;**10**(3):225-237. DOI: 10.1080/10106049.2015.1132484

[7] Shirmard H, Farahbakhsh E, Pour AB, Muslim AM, Müller RD, Chandra R. Integration of selective dimensionality reduction techniques for mineral Exploration using ASTER satellite data. Remote Sensing. 2020;**12**(8):1261. DOI: 10.3390/rs12081261

[8] Mohamed E-M, Abdellatif A, Abdellah L, Abderrazak H, Ahmed A, Kawtar B. Contribution of multispectral remote sensing to mining exploration in the Rehamna massif, Moroccan Meseta. E3S Web of Conferences. 2020;**2020**:150. DOI: 10.1051/e3sconf/20201500

[9] Feizi F, Mansuri E. Separation of alteration zones on ASTER data and integration with drainage geochemical maps in Soltanieh, Northern Iran. Open Journal of Geology. 2013;**3**(2):134-142. DOI: 10.4236/ojg.2013.32017

[10] Farrand W, Harsanyi J. Mapping the distribution of mine tailings in the Coeur d'Alene River valley, Idaho, through the use of a constrained energy minimization technique. Remote Sensing of Environment. 1997;**59**:64-76. DOI: 10.1016/S0034-4257(96)00080-6

[11] Qian D, Hsuan R, Chein-I C. A comparative study for orthogonal subspace projection and constrained energy minimization. IEEE Transactions on Geoscience and Remote Sensing. 2003;**41**:6

[12] Shih-Yu C, Chinsu L, Chia-Hui T, Shang-Ju C. Adaptive window-based Contrained energy minimization for detection of newly grown tree leaves. Remote Sensing. 2018;**10**(1):96. DOI: 10.3390/rs10010096

[13] Ren H, Du Q, Chang C-I, Jensen JO. Comparison between constrained energy

minimization based approaches for hyperspectral imagery. In: IEEE Workshop on Advances in Techniques for Analysis of Remotely Sensed Data, 2003, Greenbelt, MD, USA. 2003. pp. 244-248. DOI: 10.1109/WARSD.2003.1295199

[14] Ghulam A, Amer R. Mineral exploration and alteration zone mapping in eastern desert of Egypt using ASTER data. In: Conference Paper, ASPRS, Annual Conference, San Diego, California. 2010

[15] San BT. Hyperspectral image processing of eo-1 hyperion data for lithological and mineralogical mapping [thesis]. Geological Engineering Department, Middle East Technical University; 2008

[16] Ahmad N, Kazemi MAA, Javad A, Gholami H, Mahdavi R. Using pixel basis and subpixel based techniques to identify alteration zones (Case Study: Tange Bostanak Region). Quantitative Geomorphological Research. 2016;**5**(1):89-109

[17] Lida N, Pour AB, Askari G, Taghipour N, Pradhan B, Chang-Wook L, et al. Comparison of different algorithms to map hydrothermal alteration zones using ASTER remote sensing data for polymetallic vein-type ore exploration: Toroud-Chahshirin Magmatic Belt (TCMB), North Iran. Remote Sensing. 2019;**11**(5):495. DOI: 10.3390/rs11050495

[18] Wakila MH, Saepuloh A, Heriawan MN, Susanto A. Performance analysis of mineral mapping method to delineate mineralization zones under tropical region. IOP Conference Series: Earth and Environmental Science. 2016;**42**:012007. DOI: 10.1088/1755-1315/42/1/012007

[19] Pour AB, Hashim M, Park Y. Alteration mineral mapping in inaccessible regions using target detection algorithms to ASTER data. IOP Conference Series: Journal of Physics. 2017;**852**:012022. DOI: 10.1088/1742-6596/852/1/012022

[20] Ni L, Honggen X, Zhou X. Mineral identification and Mapping by synthesis of hyperspectral VNIR/SWIR and multispectral TIR remotely sensed data with different classifiers. IEEE Journal of Applied Earth Observations and Remote Sensing. 2020;**13**:3155-3163

Chapter 3

Geodetic Data Processing Using Soft Computing Techniques

Maan Habib

Abstract

Geodetic data processing involves using mathematical and computational techniques to analyze and interpret geospatial data about the Earth's surface and the objects and features that exist on it. This data is collected through satellite imagery, aerial photography, and surveying instruments such as total stations and GPS receivers. The applications of geodetic data processing are diverse and include land surveying, mapping, navigation, environmental monitoring, and disaster management. It is crucial to understand and manage the Earth's resources and address global challenges such as climate change, natural disasters, and urbanization. In recent years, the information technology industry has undergone a considerable transformation that has significantly impacted the development of various disciplines. Intelligent systems, powerful tools for understanding and solving complex engineering issues, have become increasingly important in this context. Soft computing techniques, including artificial neural networks, fuzzy logic, and evolutionary algorithms, are used more frequently in geodetic data processing due to their ability to handle complex, imprecise, and uncertain data. This study discusses using soft computing techniques in geodetic data processing and examines the challenges and future directions in using soft computing techniques in geodetic data processing.

Keywords: geodetic data processing, soft computing techniques, artificial neural networks, fuzzy logic, evolutionary algorithms, data fusion and integration

1. Introduction

Geodetic data processing plays a pivotal role in our ability to understand, manage, and preserve the Earth's resources. As the foundation of modern geospatial applications, it allows us to accurately represent and analyze the Earth's surface and the objects and features that exist on it. The data collected through satellite imagery, aerial photography, and surveying instruments is vital for various applications, including land surveying, mapping, navigation, environmental monitoring, and disaster management. The need for efficient and reliable geodetic data processing methods has grown significantly due to the increasing global population, rapid urbanization, and the urgency to address climate change. Accurate geodetic data is essential for informed decision-making as it helps in planning sustainable urban growth, assessing the impacts of climate change, and mitigating the effects of natural disasters.

Processing geodetic data presents several challenges, arising from the vast amounts of data collected from various sources, the inherent uncertainties and complexities associated with geospatial data, and the need for timely and accurate analysis. Traditional geodetic data processing methods rely on deterministic mathematical models and rigorous computational techniques to make sense of the data. In contrast, these methods have been effective in many cases, and they often struggle to handle the uncertainties and complexities in geodetic data. Some of the critical challenges in processing geodetic data include the following [1, 2]:

1. Data volume: The increasing availability of high-resolution satellite imagery and other geospatial data sources has led to significant growth in the volume of data that must be processed, stored, and analyzed.

2. Data quality: Geodetic data can be affected by various sources of error, including measurement errors, sensor limitations, and atmospheric disturbances. These errors introduce uncertainty and imprecision in the data, making it challenging to obtain accurate and reliable results.

3. Data integration: Geodetic data is often collected from multiple sources and in various formats, requiring effective data integration and fusion techniques to combine information and generate a coherent representation of the Earth's surface.

4. Scalability: With the growing demand for geospatial information, there is a need for scalable processing methods that can efficiently handle large-scale datasets and provide timely results.

In response to these challenges, researchers have begun to explore alternative approaches to geodetic data processing, with soft computing techniques emerging as a promising option. Soft computing is an interdisciplinary field that encompasses several computational methodologies, including artificial neural networks (ANNs) [3–5], fuzzy logic, and evolutionary algorithms (EAs). These techniques are characterized by their ability to handle imprecise, uncertain, and incomplete information, making them well-suited for addressing the challenges associated with geodetic data processing. Soft computing techniques have demonstrated their potential in various geodetic data processing tasks such as satellite image classification [6], terrain modeling [7], and geoid determination [8, 9]. They are particularly effective in pattern recognition, learning from data, and optimizing complex problems. The flexibility and adaptability of soft computing techniques have made them an attractive alternative to traditional geodetic data processing methods, often leading to improved performance and more accurate results [10, 11]. In addition, soft computing techniques are often employed in conjunction with traditional geodetic data processing methods, resulting in hybrid approaches that leverage the strengths of both paradigms [12]. This combination allows for more robust and versatile solutions capable of tackling diverse challenges in the geodetic data processing.

Accordingly, the emergence of soft computing techniques in geodetic data processing offers a promising avenue for addressing the complexities and uncertainties inherent in geospatial data [13]. As researchers continue to explore and refine these techniques, it is expected to see significant advancements in understanding and managing the planet's resources and addressing pressing global challenges such as climate change, natural disasters, and urbanization. By integrating soft computing techniques with traditional geodetic data processing methods, it can develop innovative solutions that increase the accuracy and efficiency of geodetic data processing tasks and unlock new possibilities for research and application in geospatial information. As soft computing techniques continue to gain traction in the geodetic data processing community, scholars must stay abreast of the latest developments in the field. Collaborative efforts between geodesy, remote sensing, computer science, and artificial intelligence experts will drive further innovation and progress in applying soft computing techniques to geodetic data processing. The ongoing evolution of soft computing techniques and the increasing availability of high-resolution geospatial data presents an exciting opportunity for researchers to explore new methods, tools, and applications in the geodetic data processing. As we continue to push the boundaries of what is possible with geospatial data, the integration of soft computing techniques and traditional geodetic data processing methods promises to play a pivotal role in shaping the future of the field.

2. Bibliographic review

Over the past two decades, a substantial amount of research has been conducted in the field of geodetic data processing using soft computing techniques. This bibliographic review aims to provide a comprehensive overview of the most relevant literature in this area, focusing on using artificial neural networks, fuzzy logic, and evolutionary algorithms. The Web of Science Core Collection database was used to gather the data, and VOSviewer software was employed to generate the co-words maps (**Figure 1**). Presents a co-words map, revealing the prevalence of specific soft computing techniques in geodetic data processing research over the years. The map depicts that most studies published in the early 2010s primarily focused on neural networks, reflecting the growing interest in this technique. The color scale suggests that research on other machine learning methods, such as random forest, has gained momentum in recent years, indicating diversification of soft computing techniques applied in the field.

Figure 2 provides another co-words map, highlighting the average number of citations related to specific soft computing techniques per paper. It shows that most citations are attributed to papers focused on deep convolutional neural networks. This demonstrates deep learning approaches' significant impact and recognition in the geodetic data processing. The high citation rate for deep convolutional neural networks-related papers suggests that these techniques have successfully addressed complex geodetic data processing tasks, leading to increased interest and adoption by researchers.

3. Basics of geodetic data

Geodetic data processing involves the collection, analysis, and interpretation of data about the Earth's surface and the objects and features that exist on it. This data is collected through various methods, including satellite imagery, aerial photography, and surveying instruments such as total stations and GPS receivers. Once collected, the data is processed and analyzed to extract meaningful information about the Earth's surface, which can be used in various applications, as depicted in **Figure 3**.

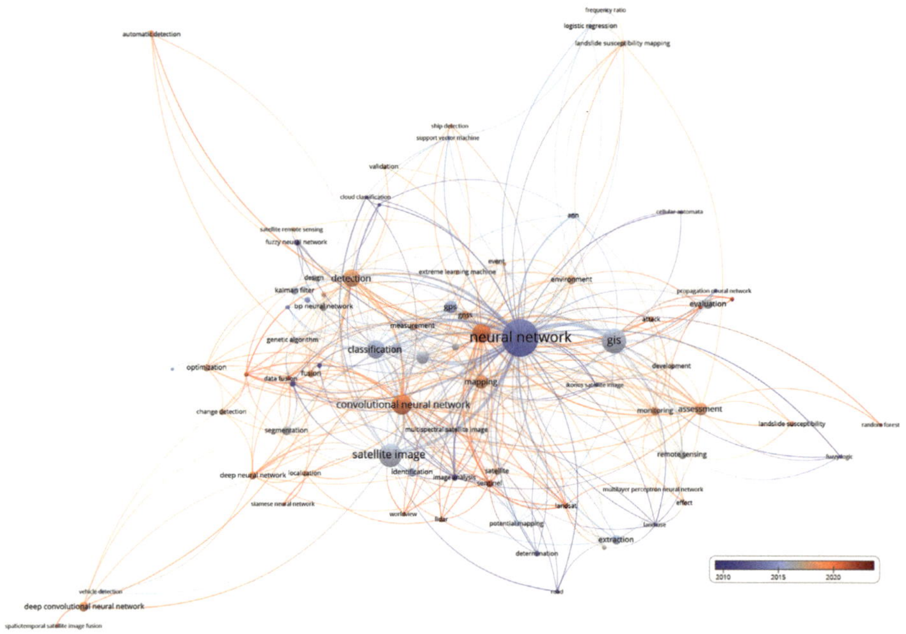

Figure 1.
Co-words map from the co-occurrence frequency of the primary keyword in papers' titles (The color scale represent years for which the majority of the paper were published).

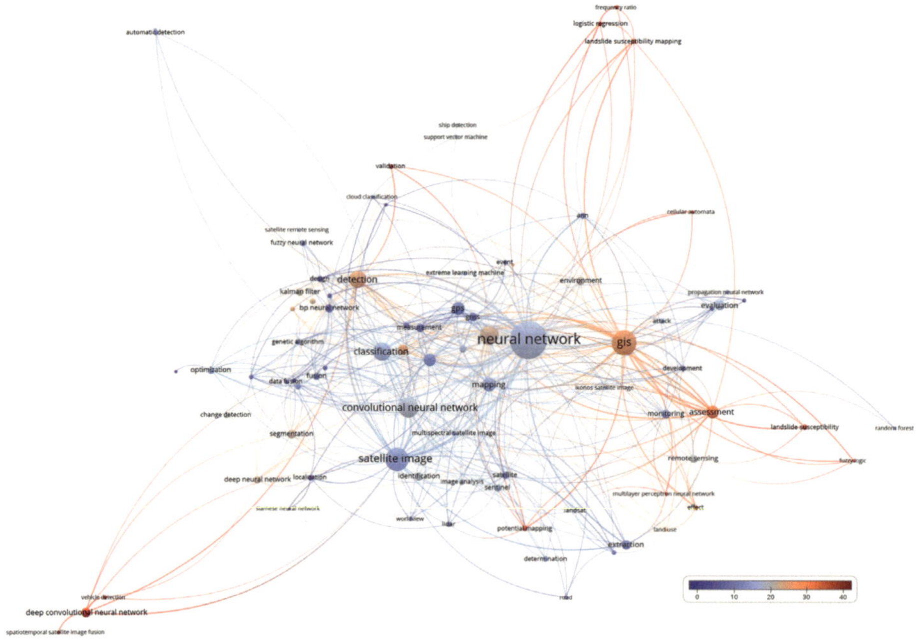

Figure 2.
Co-words map from the co-occurrence frequency of the primary keyword in papers' titles (The color scale represents the average number of citations for each of the papers published).

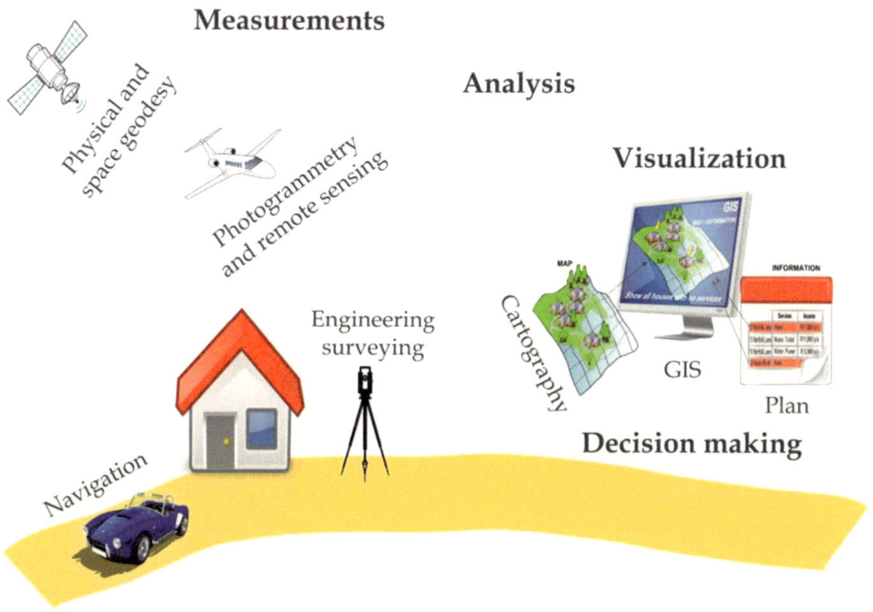

Figure 3.
Geodetic data processing.

3.1 Data collection methods

This section discusses various data collection methods for geographical analysis (**Figure 4**). They explore satellite imagery, aerial photography, and surveying instruments as essential tools for gathering accurate and reliable information about the Earth's surface [14–17].

Satellite imagery is a primary source of geodetic data and involves capturing images of the Earth's surface using remote sensing satellites. These images provide valuable information about the Earth's topography, land use, and vegetation patterns. Satellite imagery can be categorized based on spatial, spectral, and temporal resolutions. High-resolution satellite images, such as those provided by the Landsat and

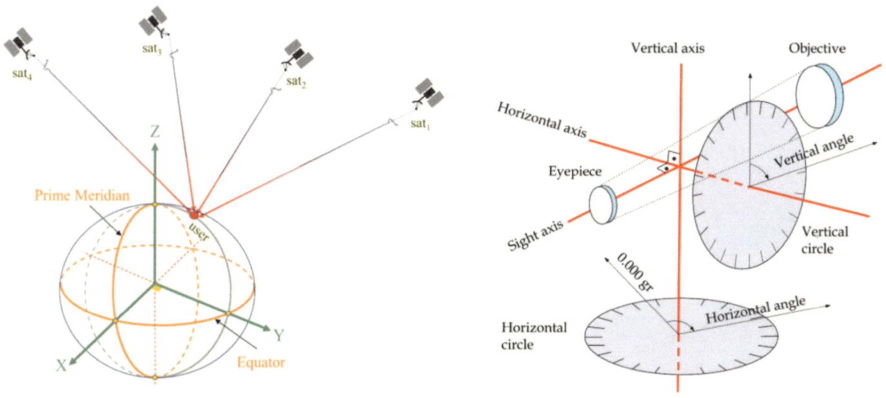

Figure 4.
Satellite and terrestrial geodetic data collection methods

Sentinel programs, allow for identifying small features and more detailed analysis. In contrast, lower-resolution images, such as those from MODIS, are helpful for large-scale environmental monitoring and global change studies.

Aerial photography involves capturing images of the Earth's surface from aircraft, balloons, or drones. Aerial photographs have been used for decades to collect geodetic data and can provide higher-resolution imagery compared to satellite imagery, allowing for a more detailed analysis of local features. Aerial photography can be either vertical, where the camera is oriented directly towards the ground, or oblique, where the camera is angled, providing a perspective view of the landscape. Advances in technology have led to the development of digital aerial cameras and photogrammetric techniques, which enable the extraction of accurate 3D information from overlapping aerial photographs.

Surveying instruments are essential tools for collecting geodetic data on the ground. Traditional instruments, such as total stations, measure angles and distances between points on the Earth's surface, allowing for the accurate determination of positions, elevations, and other spatial information. The global positioning system (GPS) and other global navigation satellite systems (GNSS) have revolutionized geodetic data collection by providing precise positioning information using signals from orbiting satellites. Other surveying instruments, such as LiDAR (Light Detection and Ranging) and ground-penetrating radar (GPR), enable the collection of high-resolution topographic and subsurface data.

3.2 Applications of geodetic data processing

This section presents the diverse applications of geodetic data processing (**Figure 5**). As a critical component in land surveying, mapping, and navigation, it provides valuable insights for effective decision-making [18, 19]. Additionally, its role in environmental monitoring and disaster management is highlighted, emphasizing its potential to facilitate sustainable development and enhance public safety [20–22].

Land surveying is a fundamental application of geodetic data processing and involves measuring and mapping the Earth's surface to establish property boundaries, rights-of-way, and other land-related features. Geodetic data processing techniques derive accurate positional information, compute areas and volumes, and generate

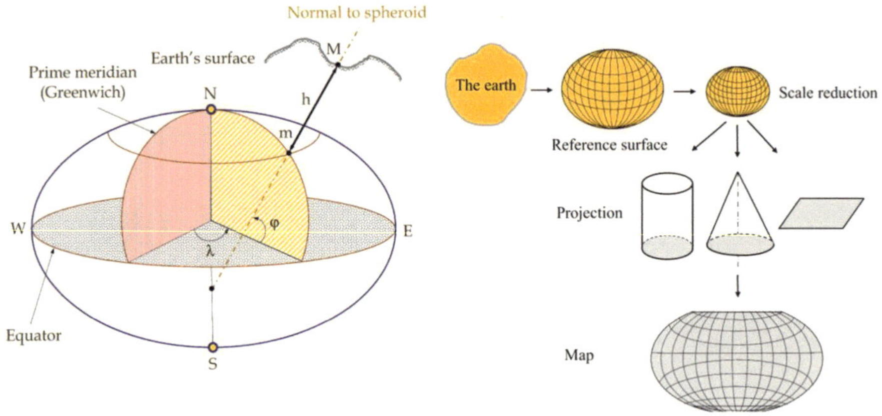

Figure 5.
Map production.

topographic maps and digital terrain models (DTMs) essential for land use planning, resource management, and property rights administration.

Mapping is a core application of geodetic data processing and includes creating various types of maps such as topographic, thematic, and cadastral maps. Geodetic data processing techniques transform raw geospatial data into useful map products that depict the spatial distribution of features such as terrain, land use, and infrastructure. Advances in geodetic data processing have led to the development of geographic information systems (GIS), allowing spatial data to be stored, analyzed, and visualized in digital form.

Navigation is an essential application of geodetic data processing, particularly in the context of GPS and other GNSS. Geodetic data processing techniques are employed to determine precise positions and velocities, compute distances and directions, and generate land, air, and sea navigation routes. The development of real-time kinematic (RTK) GPS and other high-precision positioning techniques has greatly improved the accuracy and reliability of navigation systems, with wide-ranging applications in transportation, agriculture, and other industries.

Environmental monitoring is a crucial application of geodetic data processing that involves observing and analyzing changes in the Earth's environment over time. Geodetic data processing techniques are employed to process satellite imagery, aerial photographs, and other geospatial data to assess land cover changes, deforestation, desertification, urbanization, and other environmental phenomena. Remote sensing and GIS technologies enable large-scale monitoring and analysis of environmental parameters, such as vegetation health, soil moisture, and water quality, essential for understanding the impacts of climate change and human activities on the Earth's ecosystems and resources.

Disaster management is a critical application of geodetic data processing, which plays a vital role in the assessment, mitigation, and recovery phases of natural and human-induced disasters. Geodetic data processing techniques can be utilized to analyze and model hazard scenarios, such as earthquakes, landslides, floods, and wildfires, and to generate risk and vulnerability maps that inform emergency planning and response efforts. In the aftermath of disasters, satellite imagery, aerial photography, and other geospatial data can be rapidly processed to assess the extent of damage, prioritize recovery efforts, and monitor the progress of reconstruction activities.

However, geodetic data processing is essential for various applications, from land surveying and mapping to navigation, environmental monitoring, and disaster management. The various data collection methods, including satellite imagery, aerial photography, and surveying instruments, provide diverse and complementary sources of geospatial information that can be processed and analyzed using advanced techniques and technologies. As the field of geodetic data processing continues to evolve, the development and application of innovative approaches, such as soft computing techniques, will be crucial for addressing the complex challenges associated with collecting, analyzing, and interpreting geodetic data.

4. Soft computing techniques

In recent years, the employment of soft computing techniques within geodetic data processing has garnered substantial attention as these methods have demonstrated enhanced accuracy and efficacy in addressing intricate problems [23–25]. This section delves into various soft computing techniques utilized in the geodetic data processing. They explore the applications of ANNs, fuzzy logic, EAs, and others, shedding light on their distinct advantages [26–29]. In addition, they discuss hybrid approaches that synergistically combine soft computing techniques with traditional methods, enhancing geodetic data analysis's overall efficiency and accuracy. **Table 1** lists a succinct overview of prior investigations.

No.	Reference	Soft computing technique	Remarks
1	Akar et al. [10]	Random forest	Compared random forest to classic interpolation techniques for geoid undulation prediction.
2	Akyilmaz et al. [8]	Various soft computing techniques	Investigated geoidal height transformation using soft computing methods.
3	Albayrak et al. [12]	Various soft computing techniques	Geoid modeling case study in Istanbul comparing soft computing with conventional methods.
4	Alemam et al. [3]	Artificial neural network	Improved GNSS positioning in Egypt by integrating ANN and optimal satellite configurations.
5	Beshr and Zarzoura [30]	Various soft computing techniques	Examined performance in point displacement modeling for suspension bridges using GNSS.
6	Beshr et al. [11]	Various soft computing techniques	Evaluated soft computing techniques for GNSS data processing and point displacement modeling.
7	Cakir and Yilmaz [9]	Polynomials, radial basis function, and multilayer perceptron neural network	Local geoid determination with GPS/leveling using various methods.
8	Doganalp and Selvi [31]	Polynomials, least-squares collocation, and radial basis function	Local geoid determination in strip area projects using various methods.
9	Elshambaky [32]	Artificial neural networks	Determined corrector surface for global geopotential model using GPS/leveling in Egypt.
10	Elshambaky et al. [4]	Artificial neural networks	Developed a novel three-direction datum transformation for Egyptian geodetic coordinates.
11	Garcia-Gutierrez [33]	Various soft computing techniques	Improved environmental LiDAR models for applications using soft computing approaches.
12	Kaloop et al. [34]	Machine learning approaches	Improved local geoid model accuracy using machine learning and GPS/leveling geoid height residuals.
13	Kaloop et al. [7]	Advanced soft computing techniques	Regional shoreline geoid model estimation and evaluation using advanced soft computing techniques.

No.	Reference	Soft computing technique	Remarks
14	Kaloop et al. [35]	Heuristic regression approaches	Optimized local geoid undulation model using GPS/leveling measurements and heuristic regression approaches.
15	Khurana and Saxena [6]	Various soft computing techniques	Reviewed soft computing techniques for change detection in remotely sensed images.
16	Konakoglu [36]	Artificial neural networks	Compared deep learning models for geodetic point velocity prediction.
17	Lamkai et al. [37]	Ai training algorithms and classical regression models	Developed ellipsoidal height predictive models for the Greater Kumasi Metropolitan Area Local Geodetic Reference Network.
18	Nasser et al. [38]	Convolution neural networks and empirical rules	Improved geoid accuracy using convolution neural networks and empirical rules.
19	Sorkhabi and Alizadeh [39]	Deep learning	Predicted coseismic displacement of Ahar–Varzegan earthquakes using GPS observations and deep learning.
20	Sorkhabi et al. [40]	Deep learning	Applied deep learning to GPS geodetic velocity estimation.
21	Sorkhabi et al. [41]	Deep learning	Used deep learning to predict sea-level variability and flood for coastal city resilience.
22	Thyagharajan and Vignesh [42]	Soft Computing Techniques	Reviewed soft computing techniques for land use and land cover monitoring in remote sensing.
23	Zhong et al. [43]	Computational Intelligence	Applied computational intelligence in optical remote sensing image processing.

Table 1.
Summary of soft computing techniques in the geodetic data processing.

4.1 Artificial neural networks

Artificial neural networks have been widely applied in various aspects of geodetic data processing due to their ability to model complex relationships, learn from data, and generalize from previous experiences. Some notable applications include:

a. Satellite image classification: ANNs have been employed to classify multispectral and hyperspectral satellite imagery. Compared to traditional classification methods, they have demonstrated improved accuracy in identifying land cover types, urban features, and other geospatial objects.

b. Terrain modeling: ANNs have been used for digital elevation model (DEM) generation and analysis. They can interpolate elevation data from sparse samples, detect terrain features, and model complex surface processes such as erosion and deposition.

c. Geoid determination: ANNs have been applied for geoid modeling, which is crucial for accurate height determination and geodetic positioning. They can effectively model the complex and nonlinear relationships between gravity anomalies, topography, and geoid undulations.

4.2 Fuzzy logic

Fuzzy logic has been used in geodetic data processing to address the inherent uncertainties and imprecisions in geospatial data. Some of the critical applications involve:

a. Data fusion: Fuzzy logic can be applied to fuse geospatial data from multiple sources such as satellite imagery, aerial photography, and ground-based measurements. It allows for integrating heterogeneous data while accounting for inconsistencies and uncertainties.

b. Geospatial analysis: Fuzzy logic enables the modeling of vague and imprecise spatial relationships such as proximity, similarity, and connectivity. It has been used in various geospatial analyses, including site selection, land use planning, and environmental impact assessment.

c. Uncertainty modeling: Fuzzy logic can be utilized to represent and propagate uncertainties in geodetic data processing tasks such as coordinate transformation, network adjustment, and error analysis.

4.3 Evolutionary algorithms

Evolutionary algorithms have successfully solved optimization problems in geodetic data processing thanks to their ability to search through large and complex solution spaces. Some noteworthy applications include:

a. Parameter estimation: EAs have been used for estimating model parameters in various geodetic applications such as satellite orbit determination, geoid modeling, and crustal deformation analysis.

b. Feature selection: EAs can be employed to select optimal subsets of features for geospatial data classification and analysis, resulting in reduced dimensionality and improved performance.

c. Model optimization: EAs have been utilized for optimizing geodetic models and algorithms such as geodetic network adjustment, datum transformation, and spatial interpolation.

4.4 Hybrid approaches

Hybrid approaches, which integrate soft computing techniques with traditional geodetic data processing methods, have gained popularity due to their ability to capitalize on the strengths of both approaches [13, 44, 45]. Some notable applications include:

a. Image classification: Combining ANNs and fuzzy logic has been shown to improve the classification of remote sensing images [46]. Integrating these methods can lead to more accurate and reliable results, as Mellit and Kalogirou [47] and Scott et al. [48] demonstrated.

b. Land cover classification: Mather and Tso [49] developed a hybrid approach using genetic algorithms and ANNs for land cover classification. By incorporating the optimization capabilities of EAs with the learning abilities of ANNs, their method achieved improved performance compared to individual techniques. Rabelo et al. [50] presented a hybrid approach that combines neural networks, simulation, genetic algorithms, and machine learning to address real-time sequencing and scheduling problems. The study, included in the Practical Handbook of Genetic Algorithms, highlighted the potential of integrating these advanced techniques for improved performance and adaptability when dealing with complex scheduling issues in dynamic environments.

c. Geospatial analysis: Mallick et al. [51] integrated fuzzy logic with a traditional GIS-based multi-criteria decision analysis method for landslide susceptibility mapping. Nzotcha et al. [52] proposed an integrated multi-criteria decision-making methodology for selecting optimal sites for pumped hydro-energy storage plants from a sustainable development perspective. The study combined various decision-making techniques, such as the analytic hierarchy process (AHP), fuzzy logic, and GIS-based spatial analysis, to evaluate potential locations. The integrated approach facilitated a comprehensive evaluation of environmental, social, and economic factors, leading to more sustainable and informed decision-making in site selection.

d. Spatial interpolation: Using MODIS satellite data, Kartal and Sekertekin [53] developed novel hybrid models for predicting land surface temperature (LST). Their approach combined spatial interpolation techniques of geodetic data with deep learning models, resulting in more accurate LST predictions. Jiang et al. [54] investigated the quantitative soil salinity assessment using multisource remote sensing data. They employed a hybrid approach combining support vector machines (SVMs) and ANNs to improve the accuracy of soil salinity estimation. These studies highlighted the potential of integrating advanced computational methods with traditional geospatial techniques to enhance the accuracy and reliability of environmental predictions and assessments.

These examples illustrate the potential benefits of integrating soft computing techniques (**Figure 6**) with traditional geodetic data processing methods, enabling more robust, accurate, and efficient solutions for various geospatial tasks. Combining these approaches can help address the complex challenges associated with geodetic data processing, paving the way for further advancements and innovations in the field.

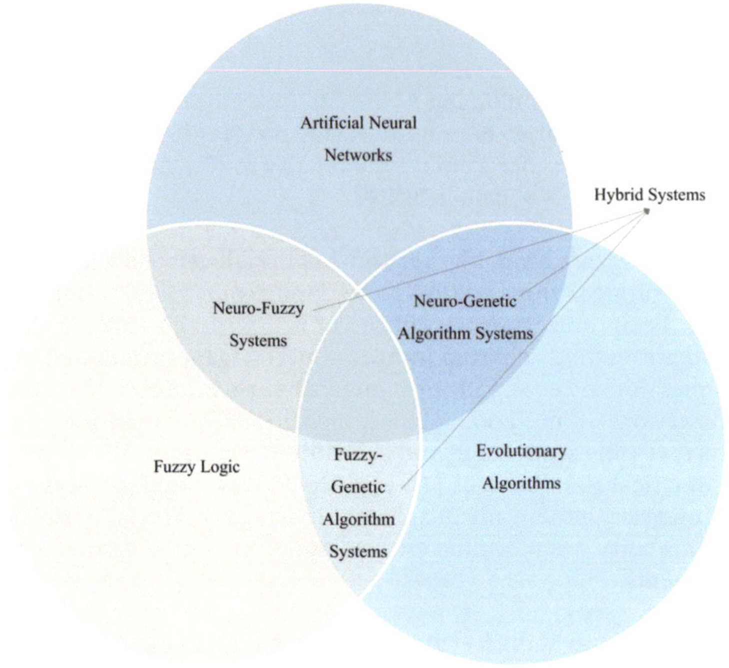

Figure 6.
Soft computing techniques.

5. Advantages and limitations of soft computing techniques

Hard and soft computing techniques (**Figure 7**) are two distinct approaches to problem-solving in computer science and artificial intelligence. Both methods have advantages and applications, and often, a combination of the two is employed to find optimal solutions [55–57]. Hard or conventional computing is rooted in mathematical

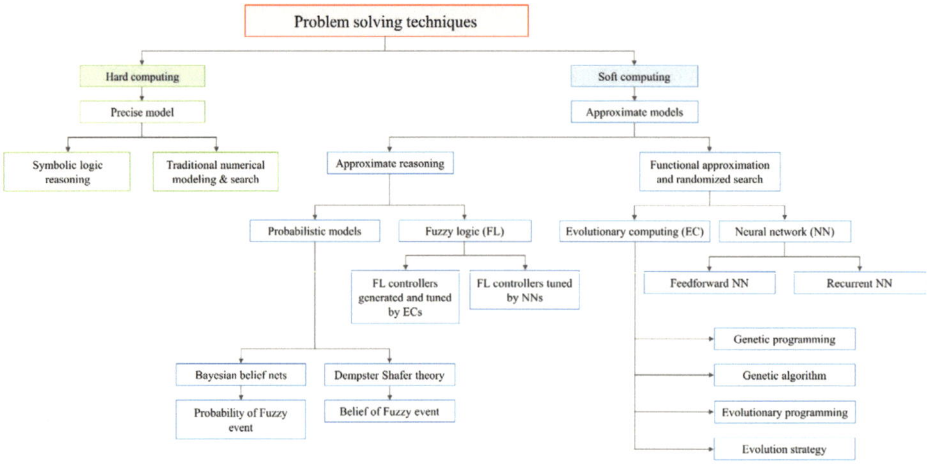

Figure 7.
Hard and soft computing techniques.

logic, precise algorithms, and deterministic models. This approach is characterized by its rigidity and strict adherence to predefined rules and instructions. It requires a well-defined problem, and its solutions are exact and reliable. In contrast, soft computing is a more flexible and adaptive approach that focuses on imprecision and uncertainty in problem-solving. Unlike hard computing, soft computing tolerates imprecise data and partial truth, making it suitable for handling complex, real-world problems [58].

The emergence of soft computing techniques has brought a new dimension to geodetic data processing, allowing for handling complex, imprecise, and uncertain data. These techniques have demonstrated their effectiveness in various geodetic data processing tasks. However, despite their potential, advantages and limitations are associated with using soft computing techniques. This section explores the benefits and challenges of using soft computing techniques in geodetic data processing, providing insight into the strengths and weaknesses of each approach [59–63]. By understanding these advantages and limitations, researchers can make informed decisions when selecting and applying soft computing techniques to geodetic data processing tasks if they comprehend these advantages and limitations.

5.1 Advantages

One of the primary advantages of soft computing techniques is their ability to handle imprecise, uncertain, and incomplete data. This is particularly important in geodetic data processing, as geospatial data often contains errors, uncertainties, and gaps due to sensor noise, measurement inaccuracies, and data loss during transmission. Soft computing techniques, such as artificial neural networks, fuzzy logic, and evolutionary algorithms, can effectively deal with these challenges by modeling and processing data flexibly and tolerantly. For example, fuzzy logic allows for representing and manipulating uncertain and imprecise data by extending traditional binary logic to incorporate the concept of partial truth. This enables the development of models and algorithms to better account for the inherent ambiguities and uncertainties in geodetic data. Similarly, artificial neural networks can learn and adapt to noisy or incomplete data by adjusting their internal weights and biases through iterative training, resulting in more robust and accurate models.

Soft computing techniques are known for their adaptability and flexibility, making them well-suited for addressing the diverse and dynamic challenges associated with geodetic data processing. Adjusting their parameters, architectures, and training strategies can easily be tailored to specific tasks and applications. In addition, soft computing techniques can often learn and adapt to new data and changing conditions, enabling them to maintain their performance even in the face of evolving problems and requirements. This adaptability and flexibility can be particularly valuable in geodetic data processing, where data volume, variety, and complexity are continuously increasing due to advancements in sensor technology and data collection methods.

5.2 Limitations

One of the main limitations of soft computing techniques is the computational cost associated with their implementation and execution. Many soft computing techniques, such as artificial neural networks and evolutionary algorithms, can be computationally intensive, mainly when dealing with large-scale datasets and complex models. This can lead to long training times, high memory requirements, and slow query response times, which can be prohibitive in some geodetic data processing scenarios, especially when real-time or near-real-time processing is required. To address this challenge, researchers have explored various strategies to improve the efficiency of soft computing techniques such as parallelization, algorithmic optimization, and hardware acceleration. However, further research and development are needed to ensure that soft computing techniques can be effectively applied to large-scale geodetic data processing tasks without compromising computational efficiency and performance.

Another limitation of soft computing techniques, particularly artificial neural networks, is their "black-box" nature, making it difficult to interpret and explain the underlying processes and decision-making. This lack of interpretability and explainability can be problematic in geodetic data processing applications, where understanding the rationale behind model predictions and decisions is crucial for validation, trust, and regulatory compliance.

Researchers have proposed various approaches to enhance the interpretability and explainability of soft computing techniques such as visualizing the internal states and activations of artificial neural networks, extracting rules from trained models, and developing hybrid models that combine soft computing techniques with more interpretable and explainable methods. Despite these efforts, further research is needed to develop more transparent and interpretable soft computing solutions for geodetic data processing, ensuring that they can be effectively integrated into decision-making processes and workflows.

6. Future directions and challenges

As geodetic data grows in volume and complexity, the need for more efficient algorithms becomes increasingly essential. Developing algorithms that can process large-scale datasets without compromising accuracy is crucial. Scholars should optimize existing soft computing techniques, such as reducing the number of iterations or simplifying model structures, to minimize computational costs. In addition, parallel and distributed computing techniques can be explored to leverage the capabilities of modern computing hardware and further enhance the efficiency of soft computing algorithms in geodetic data processing.

Fusing multiple soft computing techniques can potentially create more robust and versatile solutions for geodetic data processing tasks. For example, combining ANNs with fuzzy logic or evolutionary algorithms can lead to hybrid approaches that capitalize on the strengths of each technique while mitigating their weaknesses. Future research should investigate the optimal combinations of soft computing techniques for specific geodetic data processing tasks and develop frameworks for seamlessly integrating these approaches. Incorporating domain-specific knowledge into soft computing techniques can guide and constrain the learning and optimization processes, resulting in more accurate and interpretable models. This can be achieved by incorporating expert knowledge into the design of model structures, objective functions, or constraints. For example, geodetic data processing constraints can be integrated into the optimization process of an evolutionary algorithm to ensure the generation of feasible solutions. Future research should explore methods for effectively incorporating domain-specific knowledge into soft computing techniques and assess the impact of such integration on the performance of geodetic data processing tasks. With the increasing availability of high-resolution satellite imagery and the proliferation of geospatial data sources, there is a growing need for improved data fusion and integration techniques. With their ability to handle imprecise and uncertain data, soft computing techniques are well-positioned to address this challenge. Future research should develop novel data fusion and integration techniques that effectively combine information from diverse sources, account for uncertainties and inconsistencies, and provide a comprehensive understanding of the geospatial phenomena under investigation. Integrating soft computing techniques with big data analytics can enable the processing and analysis of massive geodetic datasets, uncovering hidden patterns and relationships that were previously unattainable. Researchers should explore methods for efficiently integrating soft computing techniques into big data analytics frameworks and assess their effectiveness in handling large-scale geodetic data processing tasks. Cloud computing offers scalable computing resources, making it an ideal platform for processing geodetic data using soft computing techniques. Future research should investigate the feasibility of deploying soft computing algorithms on cloud platforms, addressing data security, privacy, and latency challenges. The Internet of Things (IoT) can revolutionize geodetic data collection and processing by enabling real-time, continuous monitoring of geospatial phenomena. Integrating soft computing techniques with IoT devices can facilitate the development of intelligent, context-aware geospatial applications. Researchers should explore the challenges and opportunities of integrating soft computing techniques into IoT systems and develop methods for optimizing their performance in this context.

As the field of geodetic data processing continues to evolve, there is an increasing need for interdisciplinary collaboration, bringing together experts from diverse fields such as geodesy, remote sensing, computer science, and artificial intelligence. This collaboration will be critical in developing novel approaches and techniques that can effectively address the complex challenges associated with geodetic data processing. Future research should prioritize collaborative efforts, establishing partnerships and networks that facilitate sharing of knowledge, resources, and expertise across disciplinary boundaries. Developing and adopting standard protocols and benchmark datasets can help evaluate and compare different soft computing techniques in the geodetic data processing. Standardization can improve the reproducibility of research and promote the dissemination of best practices within the scientific community. Future studies should focus on establishing standardized evaluation metrics, protocols, and benchmark datasets for various geodetic data processing tasks, enabling researchers and practitioners to objectively assess the performance of different soft computing methods and identify areas for improvement. As soft computing techniques become increasingly crucial in geodetic data processing, there is a need for specialized education and training programs that equip geodesy professionals with the necessary skills and knowledge to apply these techniques in their work effectively. Universities and other educational institutions should consider developing curricula and training programs that combine geodesy, remote sensing, and geospatial sciences with artificial intelligence, machine learning, and soft computing. Additionally, short-term workshops and training courses can be organized to update existing professionals on the latest developments in the field.

With the increasing adoption of soft computing techniques in geodetic data processing, it is essential to consider the ethical implications of using these technologies. Ensuring transparency, fairness, and accountability in developing and applying soft computing models is crucial. Researchers should adhere to responsible AI principles and consider the potential societal impacts of their work, including potential biases in the data, the explainability of their models, and the environmental footprint of the computational resources required. Future studies should also explore methods for incorporating ethical considerations into designing and evaluating soft computing techniques in geodetic data processing. Developing and disseminating open-source tools and resources can play a crucial role in promoting the adoption of soft computing techniques in the geodetic data processing. Open-source tools can lower the barrier to entry for scholars who want to experiment with these techniques and foster collaboration and innovation within the scientific community. Future research should prioritize the development of open-source software, libraries, and resources for soft computing in geodetic data processing, as well as establish platforms and forums for sharing and discussing these resources.

Accordingly, the future of geodetic data processing using soft computing techniques is filled with numerous opportunities and challenges. By addressing these challenges and exploring new avenues, researchers can continue to advance the field, unlocking new possibilities and opportunities for innovation and discovery. As the integration of soft computing techniques with traditional geodetic data processing methods continues to evolve, we can expect significant advancements in our ability to understand and manage our planet's resources and address pressing global challenges such as climate change, natural disasters, and urbanization.

7. Conclusion

Soft computing techniques have demonstrated significant potential in revolutionizing the field of geodetic data processing by providing powerful and flexible tools that can effectively handle the complexities and uncertainties inherent in geospatial data. As illustrated throughout this chapter, applying artificial neural networks, fuzzy logic, and evolutionary algorithms has improved accuracy, efficiency, and adaptability in various geodetic data processing tasks, ranging from satellite image classification and terrain modeling to geoid determination and data fusion. In addition, integrating soft computing techniques with traditional geodetic data processing methods has given rise to hybrid approaches that capitalize on the strengths of both paradigms, resulting in more robust and versatile solutions. In addition, the adoption of soft computing techniques in geodetic data processing has the potential to foster greater interdisciplinary collaboration, bringing together experts from diverse fields such as geodesy, remote sensing, computer science, and artificial intelligence.

Despite the significant progress in applying soft computing techniques to geodetic data processing, challenges and limitations must be addressed to harness these benefits. Some primary concerns include the computational cost associated with specific methods, mainly when dealing with large datasets and complex models, and the interpretability and explainability of the underlying processes and decision-making. As the demand for accurate and timely geospatial information grows, researchers must continue exploring and refining soft computing techniques, addressing their limitations, and developing robust, efficient, and scalable algorithms that can be seamlessly integrated with traditional geodetic data processing methods. This requires ongoing research and innovation, as well as collaboration across disciplines, to ensure that the field remains at the cutting edge of technology and continues to meet the evolving needs of society. Future research should address the challenges and limitations of soft computing techniques and explore new avenues and opportunities for innovation. Finally, integrating soft computing techniques with other emerging technologies, such as big data analytics, cloud computing, and the Internet of Things (IoT), presents a promising avenue for future research. These interactions can unlock new possibilities for geodetic data processing by enabling real-time, large-scale analysis and decision-making and developing intelligent, context-aware geospatial applications.

Author details

Maan Habib
Faculty of Civil Engineering, Damascus University, Syria

*Address all correspondence to: maan.habib@gmail.com

References

[1] Lu Z, Qu Y, Qiao S. Geodesy: Introduction to Geodetic Datum and Geodetic Systems. United Kingdom: Springer; 2014

[2] Torge W, Müller J. Geodesy. Switzerland: de Gruyter; 2012

[3] Alemam MK, Yong B, Mohammed AS. Integration of artificial neural network and the optimal GNSS satellites' configuration for improving GNSS positioning techniques (a case study in Egypt). Artificial Satellites. 2022;**57**(1):18-46

[4] Elshambaky HT, Kaloop MR, Hu JW. A novel three-direction datum transformation of geodetic coordinates for Egypt using artificial neural network approach. Arabian Journal of Geosciences. 2018;**11**:1-14

[5] Majumdar A. Soft computing in fibrous materials engineering. Textile Progress. 2011;**43**(1):1-95

[6] Khurana M, Saxena V. Soft computing techniques for change detection in remotely sensed images: A review. arXiv preprint arXiv:1506.00768. 2015

[7] Kaloop MR, Rabah M, Hu JW, Zaki A. Using advanced soft computing techniques for regional shoreline geoid model estimation and evaluation. Marine Georesources & Geotechnology. 2018;**36**(6):688-697

[8] Akyilmaz O, Özlüdemir MT, Ayan T, Çelik RN. Soft computing methods for geoidal height transformation. Earth, Planets and Space. 2009;**61**:825-833

[9] Cakir L, Yilmaz N. Polynomials, radial basis functions and multilayer perceptron neural network methods in local geoid determination with GPS/leveling. Measurement. 2014;**57**:148-153

[10] Akar A, Konakoğlu B, Akar Ö. Prediction of geoid undulations: Random forest versus classic interpolation techniques. Concurrency and Computation: Practice and Experience. 2022;**34**(18):e7004

[11] Beshr AA, Zarzoura FH, Mazurov BT. Performance of soft computing techniques for GNSS data processing and point displacement modeling for suspension bridge. Arabian Journal of Geosciences. 2021;**14**(11):1057

[12] Albayrak M, Özlüdemir MT, Akyilmaz O, Halicioglu K. Soft computing and conventional interpolation methods in geoid modelling: A case study in Istanbul. In: JpGU-AGU Joint Meeting. United States: 2017

[13] Goodchild MF, Haining RP. GIS and spatial data analysis: Converging perspectives. Papers in Regional Science. 2004;**83**(1):363-385

[14] Blaschke T, Hay GJ. Object-based image analysis and scale-space: Theory and methods for modeling and evaluating multiscale landscape structure. International Archives of Photogrammetry and Remote Sensing. 2001;**34**(4):23-36

[15] Blewitt G, Kreemer C, Hammond WC, Gazeaux J. MIDAS robust trend estimator for accurate GPS station velocities without step detection. Journal of Geophysical Research: Solid Earth. 2016;**121**(3):2054-2068

[16] El Ghazouali S, Vissiere A, Lafon LF, Bouazizi ML, Nouira H. Optimised calibration of machine vision system for close range photogrammetry based on machine learning. Journal of King Saud University-Computer and Information Sciences. 2022;**34**(9):7406-7418

[17] Li J, Heap AD. A review of comparative studies of spatial interpolation methods in environmental sciences: Performance and impact factors. Ecological Informatics. 2011;**6**(3-4):228-241

[18] Grejner-Brzezinska DA, Toth C, Moafipoor S, Kwon J. Adaptive knowledge-based system for personal navigation in GPS-denied environments. In: Proceedings of the 2007 National Technical Meeting of the Institute of Navigation. Saudi Arabia: 2007. pp. 517-521

[19] Habib M, Farghal A, Taani A. Developing low-cost automated tool for integrating maps with GNSS satellite positioning data. Journal of Geodetic Science. 2022;**12**(1):141-153

[20] Al-Fugara AK, Pourghasemi HR, Al-Shabeeb AR, Habib M, Al-Adamat R, Al-Amoush H, et al. A comparison of machine learning models for the mapping of groundwater spring potential. Environmental Earth Sciences. 2020;**79**:1-19

[21] De Smith MJ, Goodchild MF, Longley P. Geospatial Analysis: A Comprehensive Guide to Principles, Techniques and Software Tools. Netherlands: Troubador Publishing ltd.; 2007

[22] Gorsevski PV, Gessler PE, Jankowski P. Integrating a fuzzy k-means classification and a Bayesian approach for spatial prediction of landslide hazard. Journal of Geographical Systems. 2003;**5**(3):223-251

[23] Turgut B. A back-propagation artificial neural network approach for three-dimensional coordinate transformation. Scientific Research and Essays. 2010;**5**(21):3330-3335

[24] Wang Z, Tian S. Ground object information extraction from hyperspectral remote sensing images using deep learning algorithm. Microprocessors and Microsystems. 2021;**87**:104394

[25] Ziggah YY, Youjian H, Yu X, Basommi LP. Capability of artificial neural network for forward conversion of geodetic coordinates (ϕ, λ, h) (ϕ, λ, h) to Cartesian coordinates (X, Y, Z). Mathematical Geosciences. 2016;**48**:687-721

[26] Atkinson PM, Tatnall AR. Introduction neural networks in remote sensing. International Journal of Remote Sensing. 1997;**18**(4):699-709

[27] Habib M. Quantifying topographic ruggedness using principal component analysis. Advances in Civil Engineering. 2021;**2021**:1-20

[28] Habib M, Matouk A. Integrating AHP and GIS as a decision-making tool for the optimal allocation of wind farm: A case study of Syria. In: IOP Conference Series: Materials Science and Engineering. Vol. 800. United Kingdom: IOP Publishing; 2020. p. 012019

[29] Tierra A, Dalazoana R, De Freitas S. Using an artificial neural network to improve the transformation of coordinates between classical geodetic reference frames. Computers & Geosciences. 2008;**34**(3):181-189

[30] Beshr AA, Zarzoura FH. An investigation on the performance of soft computing techniques for point displacement Modeling for suspension bridge using GNSS technique. Arabian Journal for Science and Engineering. 2021;**46**(11):10541-10555

[31] Doganalp S, Selvi HZ. Local geoid determination in strip area projects by using polynomials, least-squares collocation and radial basis functions. Measurement. 2015;**73**:429-438

[32] Elshambaky HT. Application of neural network technique to determine a corrector surface for global geopotential model using GPS/leveling measurements in Egypt. Journal of Applied Geodesy. 2018;**12**(1):29-43

[33] Garcia-Gutierrez J. Improving models for environmental applications of LiDAR: Novel approaches based on soft computing. AI Communications. 2016;**29**(1):213-214

[34] Kaloop MR, Pijush S, Rabah M, Al-Ajami H, Hu JW, Zaki A. Improving accuracy of local geoid model using machine learning approaches and residuals of GPS/levelling geoid height. Survey Review. 2022;**54**(387):505-518

[35] Kaloop MR, Zaki A, Al-Ajami H, Rabah M. Optimizing local geoid undulation model using GPS/leveling measurements and heuristic regression approaches. Survey Review. 2020;**52**(375):544-554

[36] Konakoglu B. Prediction of geodetic point velocity using MLPNN, GRNN, and RBFNN models: A comparative study. Acta Geodaetica et Geophysica. 2021;**56**(2):271-291

[37] Lamkai N, Asenso-Gyambibi D, Peprah MS, Larbi EK, Asamoah B, Okantey P. Novel ellipsoidal heights predictive models based on artificial intelligence training algorithms and classical regression models techniques: A case study in the greater Kumasi metropolitan area local geodetic reference network, Kumasi, Ghana: A case study in the greater Kumasi metropolitan area (GKMA) local geodetic reference network, Kumasi, Ghana. International Journal of Earth Sciences Knowledge and Applications. 2022;**4**(3):493-515

[38] Nasser M, Sami K, Abdalla A. Improvement of Geoid Accuracy Using Convolution Neural Networks and Empirical Rules. Turkey; 2023

[39] Sorkhabi OM, Alizadeh SMS. Coseismic displacement of Ahar–Varzegan earthquakes based on GPS observations and deep learning. Arabian Journal of Geosciences. 2021;**14**:1-6

[40] Sorkhabi OM, Alizadeh SMS, Shahdost FT, Heravi HM. Deep learning of GPS geodetic velocity. Journal of Asian Earth Sciences: X. 2022;**7**:100095

[41] Sorkhabi OM, Shadmanfar B, Al-Amidi MM. Deep learning of sea-level variability and flood for coastal city resilience. City and Environment Interactions. 2023;**17**:100098

[42] Thyagharajan KK, Vignesh T. Soft computing techniques for land use and land cover monitoring with multispectral remote sensing images: A review. Archives of Computational Methods in Engineering. 2019;**26**(2):275-301

[43] Zhong Y, Ma A, Soon Ong Y, Zhu Z, Zhang L. Computational intelligence in optical remote sensing image processing. Applied Soft Computing. 2018;**64**:75-93

[44] Atkinson PM, Tate NJ. Advances in Remote Sensing and GIS Analysis. United Kingdom: Wiley; 1999

[45] Capello R, Nijkamp P, editors. Handbook of Regional Growth and Development Theories. revised and extended second ed. United Kingdom: Edward Elgar Publishing; 2019

[46] Murmu S, Biswas S. Application of fuzzy logic and neural network in crop classification: A review. Aquatic Procedia. 2015;**4**:1203-1210

[47] Mellit A, Kalogirou S. Artificial intelligence and internet of things to improve efficacy of diagnosis and remote

sensing of solar photovoltaic systems: Challenges, recommendations and future directions. Renewable and Sustainable Energy Reviews. 2021;**143**:110889

[48] Scott GJ, Marcum RA, Davis CH, Nivin TW. Fusion of deep convolutional neural networks for land cover classification of high-resolution imagery. IEEE Geoscience and Remote Sensing Letters. 2017;**14**(9):1638-1642

[49] Mather P, Tso B. Classification Methods for Remotely Sensed Data. Germany: CRC Press; 2016

[50] Rabelo L, Jones A, Yih Y. A hybrid approach using neural networks. Simulation genetic algorithms, and machine learning for real-time sequencing and scheduling problems. In: Practical Handbook of Genetic Algorithms. CRC Press; 2019. pp. 31-52

[51] Mallick J, Singh RK, AlAwadh MA, Islam S, Khan RA, Qureshi MN. GIS-based landslide susceptibility evaluation using fuzzy-AHP multi-criteria decision-making techniques in the Abha watershed, Saudi Arabia. Environmental Earth Sciences. 2018;**77**:1-25

[52] Nzotcha U, Kenfack J, Manjia MB. Integrated multi-criteria decision making methodology for pumped hydro-energy storage plant site selection from a sustainable development perspective with an application. Renewable and Sustainable Energy Reviews. 2019;**112**:930-947

[53] Kartal S, Sekertekin A. Prediction of MODIS land surface temperature using new hybrid models based on spatial interpolation techniques and deep learning models. Environmental Science and Pollution Research. 2022;**29**(44):67115-67134

[54] Jiang H, Rusuli Y, Amuti T, He Q. Quantitative assessment of soil salinity using multi-source remote sensing data based on the support vector machine and artificial neural network. International Journal of Remote Sensing. 2019;**40**(1):284-306

[55] Das SK, Kumar A, Das B, Burnwal AP. On soft computing techniques in various areas. Comput. Sci. Inf. Technol. 2013;**3**(59):166

[56] Sinha P, Tuteja M, Saxena S. Medical image segmentation: hard and soft computing approaches. SN Applied Sciences. 2020;**2**:1-8

[57] Gupta P, Kulkarni N. An introduction of soft computing approach over hard computing. International Journal of Latest Trends in Engineering and Technology (IJLTET). 2013;**3**(1):254-258

[58] Saridakis KM, Dentsoras AJ. Soft computing in engineering design–A review. Advanced Engineering Informatics, (United kingdom). 2008;**22**(2):202-221

[59] Hajian A, Styles P. Application of Soft Computing and Intelligent Methods in Geophysics. New York, NY, USA: Springer International Publishing; 2018. pp. 25-26

[60] Ibrahim D. An overview of soft computing. Procedia Computer Science. 2016;**102**:34-38

[61] Reiterer A, Egly U, Vicovac T, Mai E, Moafipoor S, Grejner-Brzezinska DA and Toth CK. Application of Artificial Intelligence in Geodesy: A Review of Theoretical Foundations and Practical Example. United Kingdom; 2010

[62] Bernard KB, Peprah MS. Modelling local geometric geoid using soft computing and classical techniques: A case study of the University of Mines and Technology (UMaT) local geodetic

reference network. International Journal of Earth Sciences Knowledge and Applications. 2020;**2**(3):166-177

[63] Tzotsos A, Argialas D. Support vector machine classification for object-based image analysis. In: Object-based Image Analysis: Spatial Concepts for Knowledge-Driven Remote Sensing Applications. Germany: 2008. pp. 663-677

Chapter 4

Urban Trees in the Metropolitan Region of São Paulo: A Study of Geodesign and Ecosystem Services

Amanda Fruehauf, Adriana Afonso Sandre, Magda Lombardo and Paulo Renato Mesquita Pellegrino

Abstract

The Metropolitan Region of São Paulo has been presenting a scenario of high urbanization, with an increase of impermeable surfaces and verticalization to the detriment of green areas, especially urban afforestation. These scenarios contribute to the extremes of temperature and the presence of urban heat islands, especially in the more densely populated and built-up areas, where the temperature is higher than the surrounding areas that have vegetation, especially urban forestation, promoting ecosystem services such as temperature cooling. Thus, this work aimed to use Geographical Information Systems and Geodesign framework with the use of the GISColab platform, in order to study urban afforestation, as an urban forest in the study area and with specific examples, aiming at the city's socio-environmental quality. Thus, the work aims to seek nature-based solutions, such as increasing urban afforestation, in order to promote socio-environmental quality for MRSP.

Keywords: urban trees, Geodesign framework, urban ecology, nature-based solutions, socio-environmental

1. Introduction

Urban forests can be defined as a set composed of all forest fragments, parks, gardens, and street trees [1]. This network of tree canopy that comprises the urban forest, can be in urban or peri-urban areas, composed of public or private environments [2].

Its effective management can be incorporated into mitigation plans, following global agreements for climate change, such as the Paris Agreement [3] and the strategies to make cities more sustainable, expressed by the Sustainable Development Goals [4]. The urban forest is important because it provides several ecosystem services, such as recreation, walking, mitigation of climate change and improvement of air quality, public health, and protection of tangible and intangible heritage.

Ecosystem services refer to several benefits that society can obtain from ecosystems and that can be organized into four types of services: provisioning, regulating, supporting, and cultural. First, provisioning services refers to the tangible goods that ecosystems provide, such as food, fiber, and fuel. Regulating services include the processes that ecosystems provide to regulate climate, water quality, and natural hazards. Supporting services refer to the underlying processes that allow ecosystems to function, such as nutrient cycling and soil formation [5]. Finally, cultural services are the nonmaterial benefits that people obtain from ecosystems, such as spiritual and recreational values.

Considering all these types of services and their complexities is important to construct methodological strategies that are able to analyze the landscape in all aspects, aiming for its development and project planning [6]. Analyzing ecosystem services requires an interdisciplinary approach that involves ecology, economics, social science, and policy analysis. It is important to consider the multiple benefits that ecosystems provide and the trade-offs involved in managing them.

Especially on people's quality of life and health that will be directly benefited from the provision of distinct ecosystem services, promoting increased urban resilience, quality of life, and health for inhabitants [7, 8]. Overall, the improvement of the urban environment by planting trees will impact the way people live and move, which may reflect in the CO_2 emission reduction while improving mental and physical health [9].

The historical processes of land use and occupation in the Metropolitan Region of São Paulo (MRSP), exemplify the dynamics and impacts of decision-making and the consequences it has on CO_2 emissions, and number and quality of green areas in the region. Gathering more than 22 million inhabitants distributed along a 39 municipalities conurbation spread over 8051 km^2 [10], the region has two main protection areas planned to work as a rather continuous green belt across the north (Serra da Cantareira), and south (Serra do Mar) of the MRSP. Moreover, the area has an extensive reservoir system responsible for the water supply of the MRSP.

This research, which is composed of different scientific areas of specialists, aims to evaluate with Steinitz's Geodesign Framework process [11], the trend of deforestation and reforestation of MRSP until 2050. Aiming to broaden the knowledge of urban ecology, in order to pay attention to the expansion of urban afforestation, aiming at the promotion of ecosystem services, with some punctual examples included in MRSP, such as the Butantã Subprefecture, included in the Municipality of São Paulo.

Thus, the chapter discusses the importance of urban planning, aiming to increase urban afforestation as an urban forest in the study area, contributing to the socio-environmental quality of the city and its urban ecology in early, late, and non-adopter scenarios.

2. Material and methods

The Metropolitan Area of São Paulo, MRSP, is composed of 39 municipalities that cover a total area of 8051 km^2 [10] and has a population of over 22 million people (**Figure 1**).

Moreover, MRSP is one of the most populated and economically important regions in Brazil, but it also faces several environmental problems, including air pollution, water pollution—Tietê and Pinheiros rivers, which flow through the metropolitan area, are highly polluted due to untreated sewage, industrial waste, and runoff from urban areas, deforestation due to urbanization that has led to the loss of habitat and biodiversity, soil erosion, degradation of vegetation cover that also increased risk of landslides and flooding. Addressing these environmental problems requires a comprehensive approach that involves policy and regulatory frameworks, public education

DOI: http://dx.doi.org/10.5772/intechopen.1001470

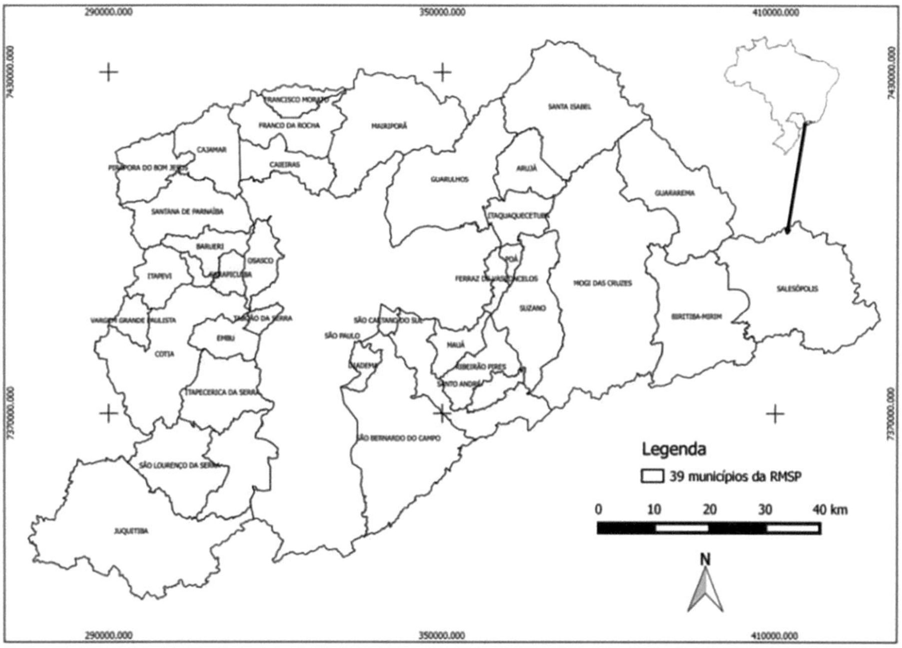

Figure 1.
Map of RMSP from the atlas of socio-environmental vulnerability of the metropolitan region of São Paulo.

and awareness, and investments in technologies to promote sustainable development. Hence, considering the objective to increase the vegetation cover and the carbon credits of at the MASP, a group of 13 researchers from various fields, worked remotely and simultaneously in a workshop using the GISColab platform[1]. This platform was developed within the scope of the Geoprocessing Laboratory of the School of Architecture of the Federal University of Minas Gerais. It has a flexible approach that is adapted to the Brazilian context, which works according to the Geodesign framework. The combination Geodesign framework and the GISColab platform allowed the codesign of ideas and plans for large regions, and are especially suitable for remote work during the pandemic.

It is emphasized that Geodesign improves the design and planning process that together with the Geographic Information System (GIS) can contribute to works that aim at the sustainability of the area and meet the population's goal of environmental quality [12]. The Geodesign process typically involves several stages, including problem identification, data analysis and modeling, design alternatives, evaluation and selection of the best alternative, and implementation and monitoring. In this sense, Geodesign also involves the iterative design and feedback loops to refine, and improve the solutions. And it has been used to promote and plan intelligent landscapes. In a nutshell, promoting and planning intelligent landscapes involves designing and managing landscapes that are sustainable, resilient, and responsive to the needs of people and the environment. Geodesign process is open for the integration of diverse

[1] The Brazilian Geodesign platform - GISColab has the premise of supporting and connecting Spatial Data Infrastructure (SDI), WebGis, and Geodesign in order to provide integrated and georeferenced information on the territory, enabling a wide availability of data to support the discussions and negotiations on the adopted spatial profile [Locosselli]. Available at: <GisColab: http://www.giscolab.com/geodesign/#/

stakeholders, including local communities, government agencies, and businesses, to develop strategies that balance competing priorities and promote sustainable development.

So that green spaces can be relevant to support the bodies and decisions of urban planning, aiming at the conservation and management of spaces [13].

It is necessary to develop viable alternatives for the implementation of green areas in the cities, where it is necessary to think about their distribution in public and private facilities, distribution in the cities and among the communities, covering their occupation in the use of the soil, making their application viable, escaping from the onerous pattern and longtime demand of the cost involved that has been used [14].

In this way, the use of Geographic Information System (GIS), can assist in landscape planning, aiming to increase urban forestation. GIS presents a flexible environment for input, analysis, and mapping of data for land use and land cover, for example Ref. [15].

Initially, the available database on the GISColab platform was analyzed by the participants, who identified the most relevant layers to the MRSP case and highlighted essential data, which had not been covered by the national project organization, such as layers of urban heat island and flooding prone areas, available at the MASP's Socio-Environmental Vulnerability Atlas, and information about contaminated areas, landfills, and postindustrial areas, where increased afforestation can be of especial benefit (**Figure 2**).

With the main objective of achieving the objective of establish an urban forest, improving the street trees, water managing, and natural and planted forest areas, decreasing the phenome of urban heat islands in the early, late, and non-adopter scenarios (**Table 1**).

About the late-adopter (2035) and non-adopter (2050) the values of increasing or decreasing percentages related to 2020/21 were estimated for 2035 and 2050 by evaluating the linear trends from historical datasets starting in 1985 up to the present. Finally, the information obtained was compared with 2020/21 in relative terms. In

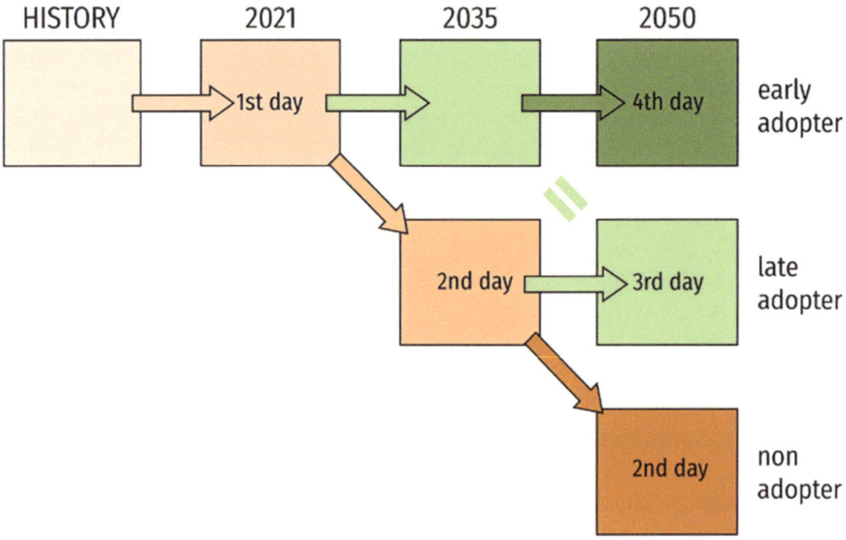

Figure 2.
Geodesign workshops workflow. Source: Authors. Adapted from IGC, 2021.

	Current	**Early**		**Late**		**Non**	**References**
	2020	**2035**	**2050**	**2035**	**2050**	**2050**	
Number of street trees based on the balance between planted and fallen plus removed trees.	652,146	30%	50%	9%	30%	18%	[16, 17]
Impacts on the recharge of water reservoirs due to expansion of impervious cover.	1.132. 610.000 m^3	10%	25%	−38%	−10%	−76%	[18, 19]
Detention ponds over urban forests to be built..	48 un.	−4%	−8%	73%	83%	146%	[20]
Green detention reservoirs and recovery wetlands to be built..	0 un.	35%	70%	1%	10%	1%	[12, 17]
Increase/Reduction in informal settlements.	88,34 km^2	0%	−5%	2.5%	0.5%	5%	[10, 21]
Urban sprawl evolution	2.038,69 km^2	8%	4%	17%	9%	34%	MapBiomas platform Growth compared with 1985–2019 period
Increase/Reduction in natural forest areas due to urban sprawl/reduction.	3.581,85 km^2	15%	30%	−2%	−2.5%	−4%	
Increase in planted forest (Eucalyptus, Pinus and Araucaria) areas due to urban sprawl/reduction.	260,92 km^2	10%	15%	40%	45%	78%	
Increase in the urban heat islands: average of the surface temperature of central area of SP less rural areas.	15 °C	10°C	5°C	18°C	10°C	22 °C	[21]

Adapted from IGC, 2021.

Table 1.
Predicted changes of the metropolitan region of São Paulo.

order to predict urban sprawl, two main sources of information were used. The census [22] provided an overview of the slums increment and distribution until 2010, in addition to the data available on MapBiomas for the evaluation of natural and planted forest and urban settlement areas (MapBiomas/IBGE)[2]. For street trees, data from the

[2] The Expressive Vegetation Map in GISColab was based on Vegetable Cover Maps of Brazilian Biomes (http://mapas.mma.gov.br/mapas/aplic/probio/datadownload.htm?/mosaicos_vegetacao/) and developed by GeoPrea Laboratory (EAU/UFMG).

Municipal Plan for Urban Trees [23] were used, including data on tree fall, replacement of senescent individuals, and planting. For the Late adopter (2050) and Early adopter (2035 and 2050) changes in tree cover are presented in percentages related to 2020/21. For these projections, goals for afforestation of the MRSP were discussed and defined among the different researchers during the workshop.

3. Results and discussions

In order to obtain an increase in street afforestation, a study was made aiming to increase this in the Butantã Subprefecture, located in the Municipality of São Paulo that is in the MRSP (**Figure 3**). The GIS resource was used, with the use of the free software QGIS, proposing a geodesign map (**Figure 4**).

There are in this Subprefecture both high-standard areas, mainly in Morumbi, from middle-class neighborhoods to slums. But in general, it is a very expressive population of medium to high socio-economic status [24].

This geodesign-based map is the intersection between the 2013 afforestation map and the proposed afforestation map. It demonstrates the proposed geodesign for the study area as a nature-based solution. Thus, one can visualize the green infrastructure gain with emphasis on urban road forestation in the Sub-municipality of Butantã, obtaining a total of 54,469 mapped trees [25].

This work showed that the Butantã Subprefecture, in 2013, had 33% of tree canopy, with this proposal to implement the tree canopy, the green area is increased by 3%, totaling 36% of green infrastructure with emphasis on tree canopy, which represents an amenity of 1.5°C, as follows (**Table 2**).

The workshop that was made took place remotely, which take place to use the GIS and Geodesign approach, having the main objective to increase the vegetal cover in urban, peri-urban, and rural areas, in public and private property, and they were related mainly to the conservation of protected areas and existing vegetation, the ecological restoration of riparian forests, and adequate management of the urban afforestation and planning guidelines for the new green areas to be proposed in the MRSP.

For the measurement of various data, such as rainfall density or quantification of urban forestation, geospatial technology, such as Remote Sensing (RS), Geographical

Figure 3.
Location of the Butantã subprefecture in the municipality of São Paulo.

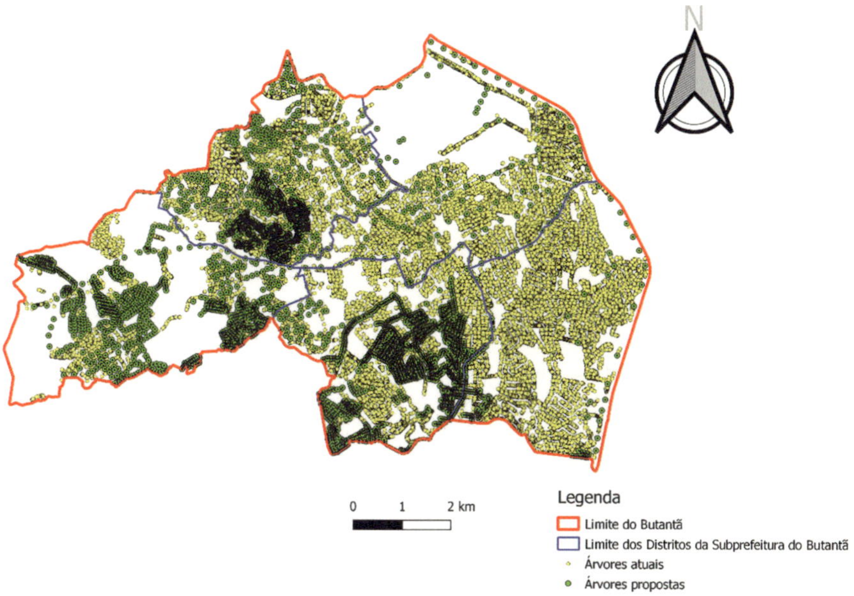

Figure 4.
Map of Butantã, with a proposal for increasing urban forestation.

Period	Total tree canopy area	Percentage of green infrastructure	Medium temperature
2013	10,18 km^2	33%	24.86°C
Proposal	11,80 km^2	36%	23.36°C

Table 2.
Status of the Butantã Subprefecture in total area of tree canopy, percentage of Green Infrastructure and average temperature in 2013 and the change with the proposal to implement the tree canopy [25].

Information Systems (GIS) and Global Positioning System (GPS) has been advancing and prevailing in many terrain space and urban planning surveys, aiming for accurate, continuous, and highly precise results [12].

Geodesign provides a framework of methods and technologies for building skills and indicators. Being an improved technological method that merges the rigid form with the planning proposal with impact simulations that occur in the geographic context [22].

As examples of these strategies, one can mention the support to plant trees in particular areas through public policies and environmental compensation mainly for the construction industry, the support to plant along avenues, and the transformation of MRSP's roads landscapes.

It is necessary for the importance of implementing afforestation in the analysis of the landscape, respecting the species with their ideal planting location, when there is a drastic intervention with irrigation and not attention to the planting location, demonstrating that the species are not in ideal locations for their typical climatic conditions. On the other hand, commercial forestry trials are established to evaluate the suitability of species and provenance [26].

The water supply is also prone to the effects of increasingly frequent climate extremes, such as the drought caused by an unusual mid-troposphere blocking, during the summer of 2014 [27]. Due to the observed values of the total volume of water in the reservoirs of the MRSP [28], very low in 2005, 2014, and 2015, also the downward trend from 2020 and the forecasts of specialists estimates there is a tendency to reduce the volume of water available for supply. At the end of 2014, the system had a negative 2% total volume compared to normal.

With respect to the theme of water reservoirs, **Figure 2** illustrates a trend toward a reduction in the percentage of total volumes, starting in the years 2020–2021. The years 2014 and 2015 were critical for the supply system, characterizing a real water crisis at MASP. The calculations made from the equation obtained from the trend line of the graph in **Figure 5** showed a reduction in the percentage of the total volume of the reservoirs, as shown in **Table 1**.

Moreover, it is important to point out that in this scenario, significant environmental, social, and economic changes will occur within and outside cities as they grow in size and complexity, economic losses and poor livelihood are also a result of the higher vulnerability of cities to extreme precipitation events [29].

In order to achieve the goal of increasing urban forestation in the study area, one must think about the change that can be obtained in the city from 2035 to 2050 (**Figure 6**). In order to occupy the land use in a balanced way that promotes the quality of life of the inhabitants and the main focus of this work is the increase of urban forestation. Nature-based solution is widely used to promote socio-environmental quality.

3.1 Nature-based solution (NbS)

The NbS arise as a way to use environmental sciences to solve problems such as mitigating climate change, improving sustainable livelihoods, and protecting natural ecosystems and their biodiversity [14].

The importance of adapting buildings and urban areas, the implementation of green areas, as a means of increasing the humidity of the area, and being proactive urban [27].

Trees are essential elements in nature-based solutions, as they simultaneously promote quality of life and biodiversity while providing mitigation and adaptation ecosystem services in cities [6].

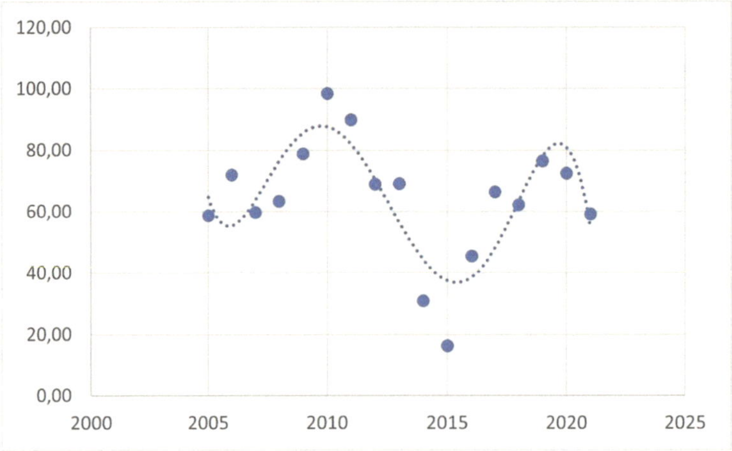

Figure 5.
Percentage of the total volume of MASP reservoirs in relation to the year.

Figure 6.
Spatial distribution of the most relevant layers of the metropolitan region of São Paulo 1. Adapted from IGC, 2021.

Therefore, the focus in research about urban planning should be on central areas that have a large accumulation of vacant buildings.

These policies should provide structure for urban redevelopment projects through urban design with nature-based solutions and green infrastructure, in a wide participatory and inclusive manner, by maintaining the population in the site, but within a secure and eco-friendly condition. In addition, one must also consider the relationship between trees and urban water.

In the scenario of 2050, the main proposal was to increase tree coverage in the built area, then the group discuss real examples of sustainability and resilience cities, such as Singapore, and also the revision of local experiences such as Maringá city, Brazil. They are public policies that encourage urban planting and afforestation, considering its ecosystem services; redesigning existing streets with nature-based solutions and focus on the implementation of urban forestry.

Thus, to achieve a resilient urban ecology, it is necessary to think of the landscape as an interdisciplinary field, with heterogeneous elements, without ignoring its natural rhythm of intensity and disturbances. Environmental management of public policy can adopt the adaptive concept, where there is room to work out the uncertainties of the natural and urban environment and learn in the process, in order to improve management [26].

Resilience requires, that there is urban landscape planning, identifying the disturbances that it has and seeking solutions, of how the study area can adapt to respond to the disturbances that the urban environment can generate in nature that affect the socio-environmental quality of life [30].

Sustainability meets NbS, which can be applied in urban and landscape planning, provided that the contrasts of the urban area and its fluidity are understood. Thus, important that resilience theory is applied, aiming for a solution to the challenges of sustainability [26].

Also, the application of NbS needs stakeholders and decision-makers to find reliable alternatives to usual monofunctional heavy engineering solutions that are becoming inadequate to cope with the complex environmental, social, and economic challenges of cities [31].

4. Conclusion

This chapter discussed how urban trees can provide a wide range of ecosystem services that benefit both the environment and people living in MRSP. Some examples of ecosystem service provided by urban trees are the improvement of air quality by absorbing pollutants, such as nitrogen oxides and particulate matter, which can have negative health impacts, carbon sequestration, and temperature regulation—which can help to reduce the urban heat island effect and lower temperatures in urban areas, stormwater management and biodiversity conservation.

The Geodesign framework, following Stein's framework with the GISColab platform, assisted in the dynamics and integration of the landscape aimed at understanding the study area, visualizing the challenges, and pursuing NbS.

According to the scenarios, it was observed if no measures are taken for changes that favor natural resources in the landscape, environmental degradation, increased urban heat island and reduced water supply. By 2035, the trends would be similar. However, if adaptations in urban planning were made, innovative technologies would be used, aiming to implement greater afforestation and in the search for NbS for the MRSP.

It is noteworthy that the work with GISColab allied with GIS, was important at the time when the work took over, due to the pandemic situation can be done remotely and unite several researchers to collaborate in the elaboration of different scenarios of the urban canopy of MRSP.

Finally, this chapter also argued—in an exploratory fashion—about how to promote and plan intelligent landscapes, such as addressing the environmental problems MRSP. Important process is to identify the specific environmental problems facing the MRSP to design and plan that integrate ecological, social, and economic considerations, such as implementing green infrastructure, promoting sustainable transportation, and creating green spaces for public use. Overall, in the case study, urban trees play a vital role in providing ecosystem services in MRSP that benefit both the environment and people living in urban areas. By planting and maintaining urban trees, São Paulo can promote a more livable, healthy, and sustainable neighborhood.

Thanks

The authors would like to thank Ana Clara Moura (UFMG) who coordinated the project "Geodesign Brasil: Trees for Metropolitan Regions;" for the support of the Geoprocessing Laboratory (Geoproea/EAUFMG) staff; and the GISColab platform (CNPq Project 401,066/2016–9/FAPEMIG PPM-00368-18). And this work was carried out with support from the Coordenação de Aperfeiçoamento de Pessoal de Nível Superior—Brazil (CAPES)—Funding Code 001.

Author details

Amanda Fruehauf[1]*, Adriana Afonso Sandre[2], Magda Lombardo[1] and Paulo Renato Mesquita Pellegrino[2]

1 University of Sao Paulo, Piracicaba, Brazil

2 University of Sao Paulo, São Paulo, Brazil

*Address all correspondence to: amandalombardo@usp.br

References

[1] da Silva Filho DF et al. Urban forest indicators from high resolution multispectral aerial imagery. Scientia Forestalis/Forest Sciences. 2005:88-100

[2] Torres H, Alves H, Oliveira MA. São Paulo peri-urban dynamics: some social causes and environmental consequences. Environment & Urbanization. 2007;**19**(1):207-223. DOI: 10.1177/0956247807076784

[3] Sant'Anna CG. A infraestrutura verde e sua contribuição para o desenho da paisagem da cidade [thesis]. Brasília: Universidade de Brasília; 2020

[4] Botton GZ et al. As construções das abordagens conceituais de cidades sustentáveis e inteligentes para superar os desafios dos objetivos do desenvolvimento sustentável. Desafio Online. 2021;**9**(3)

[5] Locosselli GM, Miyahara AL, Cerqueira P, Buckeridge MS. Climate drivers of tree fall on the streets of São Paulo, Brazil. Trees. 2021:1-9. DOI: 10.1007/s00468-021-02145-4

[6] Maruyama CC, Fruehauf AL, Lombardo MA. Infraestrutura Verde para mitigação de Ilha de Calor e sequestro de carbono da Região Metropolitana de São Paulo. In: Fruehauf AL, Rosa AA, Maruyama CC, Coelho MA, editors. Geodesign no Brasil: abordagens para o planejamento ambiental urbano. São Carlos, SP: Pedro & João Editores; 2022

[7] Gonçalves JM, Gama JMRF. A systematization of policies and programs focused on informal urban settlements: Reviewing the cases of São Paulo, Luanda and Instambul. Journal of Urbanism. 2020;**13**(4):466-488. DOI: 10.1080/17549175.2020.1753228

[8] Gómez-Baggethun E, Barton DN. Classifying and valuing ecosystem services for urban planning. Ecological Economics. 2013;**86**:235-245

[9] Konijnendjik CC. A decade of urban forestry in Europe. Forest Policy and Economics. 2003;**5**(2):173-186. DOI: 10.1016/S1389-9341(03)00023-6

[10] Instituto Brasileiro de Geografia e Estatística. Censo Brasileiro de 2010. Rio de Janeiro: IBGE; 2012

[11] Steinitz C. A Framework for Geodesign: Changing Geography by Design. Redlands: ESRI Press; 2012

[12] Brito FM, Miralgia SGEK, Semensatto DL Jr. Ecosystem services of the Guarapiranga reservoir watershed (São Paulo, Brazil): Value of water supply and implications for management strategies. International Journal of Urban Sustainable Development. 2018;**10**(1):49-59. DOI: 10.1080/19463138.2018.1442336

[13] de Almeida ALBSS. The Value of Trees: Trees and Urban Forest in Lisbon [Thesis]. Lisbon, Portugal: Universidade Técnica de Lisboa Instituto Superior de Agronomia; 2006

[14] Cohen-Shacham E et al. Nature-Based Solutions to Address Global Societal Challenges. Vol. 97. Gland, Switzerland: IUCN; 2016

[15] Liu T, Yang X. Monitoring land changes in an urban area using satellite imagery, GIS and landscape metrics. Applied Geography. 2015;**56**:42-54

[16] São Paulo (Prefeitura), Fundação Centro Tecnológico de Hidráulica (Org.). Caderno de bacia hidrográfica: córrego Jaguaré. São Paulo: Siurb/FCTH; 2016

[17] Sabesp, Companhia de Saneamento Básico do Estado de São Paulo: Portal dos mananciais. Available from: http://mananciais.sabesp.com.br/Situacao [Accessed: April 29, 2021]

[18] Tundisi JG. Ciclo hidrológico e gerenciamento integrado. Cienc. Cult. 2003;**55**(4):31-33. DOI: 10.1177/0956247807076784

[19] São Paulo (Prefeitura). Plano Diretor de Macrodrenagem da Bacia do Alto Tietê. São Paulo, SP: Departamento de Águas e Energia Elétrica; 2013

[20] Pasternak S, Bogus LM. Favelas na Macrometropole Paulista. In: A Seminario Internacional de Investigación en Urbanismo. XII Seminario Internacional de Investigación en Urbanismo, São Paulo-Lisboa, 2020. São Paulo: Faculdade de Arquitetura da Universidade de Lisboa; 2020. DOI: 10.5821/siiu.10116

[21] Lombardo MA. Ouso de geotecnologias na análise das mudanças climáticas na metrópole de São Paulo. Revista Geográfica de América Central. 2011;**2**:1-19

[22] Dangermond J. Geodesign and GIS–designing our futures. Proceedings of Digital Landscape Architecture. 2010:502-514

[23] Moreira TCL, Polizel JL, Santos IS, Filho DFS, Bensenor I, Lotufo PA, et al. Green spaces, land cover, street trees and hypertension in the megacity of São Paulo. International Journal of Environmental Research and Public Health. 2020;**17**(3):725. DOI: 10.3390/ijerph17030725

[24] Morato RG, Kawakubo FS. Análise espacial da desigualdade ambiental na subprefeitura de Butantã, São Paulo-SP. Revista brasileira de geografia médica e da saúde. 2007:66-73

[25] Fruehauf AL. Analysis of Land Use, Thermal Field and Vegetation Index for the Implementation of a Green Infrastructure in the Search for Urban Environmental Quality in the Butantã Subprefecture, in the Municipality of São Paulo-SP [Thesis]. Piracicaba, Brazil: Luiz de Queiroz College of Agriculture, University of São Paulo; 2020. DOI: 10.11606/D.11.2020.tde-01102020-171031

[26] Ahern J. From fail-safe to safe-to-fail: Sustainability and resilience in the new urban world. Landscape and Urban Planning. 2011;**100**(4):341-343

[27] Marengo JA, Alves LM, Ambrizzi T, Young A, Barreto NJC, Ramos AM. Trends in extreme rainfall and hydrogeometereological disasters in the metropolitan area of São Paulo: A review. Annals of the New York Academy of Sciences. 2020;**1472**(1):5-20. DOI: 10.1111/nyas.14307

[28] Mcpherson EG, Simpson JR, Peper PJ, Xiao Q. Benefit-cost analysis of Modesto's municipal urban forest. Journal of Arboriculture. 1999;**25**(5):235-248

[29] Emplasa, Empresa Paulista de Planejamento Metropolitano S/A. Biblioteca Virtual. São Paulo, SP, Brasil; 2022

[30] Vale L, Campanella J, Thomas J. The Resilient City: How Modern Cities Recover from Disaster. Oxford, UK: Oxford University Press; 2005

[31] Bélanger P. Landscape Infrastructure, Urbanism beyond Engineering [Thesis]. Wageningen, Netherlands: Wageningen University; 2013

Chapter 5

Modeling Invasive *Prosopis juliflora* Distribution Using the Newly Launched Ethiopian Remote Sensing Satellite-1 (ETRSS-1) in the Lower Awash River Basin, Ethiopia

Nurhussen Ahmed and Worku Zewdie

Abstract

Ethiopia successfully launched its first earth-observing satellite sensor in December 2019 for the purpose to manage natural resources and enhance agriculture. This study aimed at evaluating the potential of Ethiopian Remote Sensing Satellite 1 (ETRSS-1), for the first time, for detecting and mapping *Prosopis juliflora* distribution. To better test its potential, a comparison was made against the novel Sentinel-2 Multispectral Instrument and Landsat-8 Operational Land Manager datasets. Radiometric indices (Scenario-1) and spectral bands (Scenario-2) derived from these sensors were used to model the distribution of *Prosopis juliflora* using the random forest modeling approach. A total of 241 georeferenced field data on species presence and absence data were used to train and validate datasets in both scenarios. True skill statistics (TSS), area under the curve (AUC), correlation, sensitivity, and specificity were used to evaluate their performance. Our results described that the ETRSS-1-derived variables can be sufficient for modeling and mapping of *P. juliflora* distribution in such settings. However, higher performance was found from Sentinel-2 with AUC > 0.97 and TSS > 0.89, and followed by Landsat-8 with AUC > 0.93 and TSS > 0.77 and ETRSS-1 with AUC > 0.81 and TSS > 0.57. The lower performance of ETRSS-1 compared to Landsat-8 and Sentinel-2 datasets, however, is partly due to its coarse spectral resolution. Hence, improving the spectral resolution of ETRSS-1 might increase its accuracy.

Keywords: Ethiopian remote sensing satellite 1, Landsat-8, Sentinel-2, *Prosopis juliflora*, random forest

1. Introduction

Since the successful launch of the first satellite in 1957, Sputnik-1, satellite remote sensing has been in continuous development [1, 2]. Nowadays, several state-of-the-art sensors are on board providing multispectral and hyperspectral information. After the successful launch of Landsat-1 in July 1972, the joint National Aeronautics and Space Administration (NASA) and the United States Geological Survey (USGS) launched Landsat Thematic Mapper (TM) in 1982, Landsat Enhanced Thematic Mapper (ETM) in 1999, and Landsat Operational Land Manager (OLI) in 2013 [3]. Moreover, NASA is still under continuous development and launched Landsat-9 (OLI-2) in September 2021. In addition, the French government has successfully launched seven Satellite Pourl' Observation de la Terre (SPOT) since its first successful launch, SPOT-1, on 22 February 1986. Moreover, the European Space Agency (ESA) Copernicus program launched a series of satellites of Sentinel-1A and 1B, Sentinel-2A and 2B, and Sentinel-3A and 3B. In addition, ESA will launch Sentinel-4, Sentinel-5, and Sentinel-6 and another 6 second-generation satellite series (Sentinel-7 to 12) in the coming years.

Though many satellite sensors are currently available onboard, these satellites are mainly launched and owned by developed nations. Due to its high investment and skilled manpower requirement, so far, only 11 African nations have successfully launched 36 satellites into orbit with the collaboration of developed nations [2]. In collaboration with the Chinese government, Ethiopia successfully launched its first (Ethiopian Remote Sensing Satellite (ETRSS-1)) satellite in December 2019. It is a sun-synchronous orbit providing information at 13.75 meters of spatial resolution every 4 days. Beside, they are multispectral imagery providing information at four different spectral bands (Blue, Green, Red, and Near-Infrared). The launch of these and other states of art earth-observing satellite sensors is believed to provide reliable information for natural resource management [4, 5].

Remotely sensed-based modeling and quantifying of invasive species distribution can benefit from remotely sensed derived radiometric indices, and/or spectral bands, and/or environmental variables and/or their combinations. The commonly used approach is the use of remotely sensed derived vegetation, soil and water indices, and biophysical variables such as normalized difference vegetation index (NDVI), soil adjusted vegetation index (SAVI), normalized difference water index (NDWI), and leaf area index (LAI). A study by Ahmed et al. [6] and Ng et al. [7] employed vegetation indices (Vis) derived from Sentinel-2 to model the distribution of invasive *Prosopis juliflora* (*P. juliflora*) in Ethiopia and Kenya, respectively. Beside, Wakie et al. [8] used VIs derived from Moderate Resolution Imaging Spectroradiometer (MODIS) to map and model the distribution of *P. juliflora* in Ethiopia. Moreover, detecting and mapping invasive species can also be possible using spectral bands. A study by Arogoundade et al. [9] successfully mapped the distribution of *Parthenium hysterophorus* invasion using Sentinel-2 Multispectral Instrument (MSI) derived VIs and spectral bands. Furthermore, several studies also employed the integration of radiometric indices, spectral bands, and topo-climatic variables for mapping invasive species distribution [9–11].

Owing to the spectral, spatial, and temporal variation of Earth Observation (EO) satellite sensors, evaluating the relative potential of models derived from varied EO datasets is highly required. Models based on the freely available Sentinel-2 MSI and Landsat-8 OLI, for example, were evaluated for different applications such as soil salinity detection [12], geological mapping [13], *water hyacinth* [14], and greenhouse gas detection [15]. A study by Jensen et al. [16] evaluated the potential of models derived from Sentinel-2 variables against Airborne Visible/Infrared Imaging Spectrometer (AVIRIS) hyperspectral data for mapping invasive *Kudzu* species distribution in the USA. Similarly, a study by Ng et al. [7] made a comparison between models derived from Sentinel-2 and Pléiades-derived for mapping invasive *P. juliflora* in Kenya. In addition, Alvarez-taboada et al. [17] used unmanned areal vehicle (UAV) and WorldView-2-derived variables to map invasive species distribution. Moreover, previous studies in the study area were focused on either MODIS [8], Landsat-8 OLI [18], or Sentinel-2 [6] derived variables. However, identification and mapping of invasive *P. juliflora* distribution using ETRSS-1 are missing. Hence, this study aims, for the first time, at evaluating the potential of the ETRSS-1-derived to map and model the distribution of invasive *P. juliflora* in arid and semiarid regions of Ethiopia. To better evaluate its performance, a comparison was made against Landsat-8 OLI and Sentinel-2-derived variables.

2. Materials and methods

This study evaluated the relative potential of models derived from ETRSS-1 for detecting and mapping the distribution of invasive *P. juliflora* in the lower Awash River basin, Ethiopia. Its potential was also evaluated against Landsat-8 OLI and Sentinel-2B-derived variables. The overall process to evaluate its potential is presented in **Figure 1**.

2.1 Study area and species

The study was carried out in the lower Awash River basin, Ethiopia, extending from 40.71° to 41.1° longitude and 11.41° to 12.26° latitude (**Figure 2**). It covers 3381.3 KM^2 ranging between 240 and 1341 meters elevation above sea level. According to the National Metrological Agency (NMA) (2020), the mean annual rainfall and

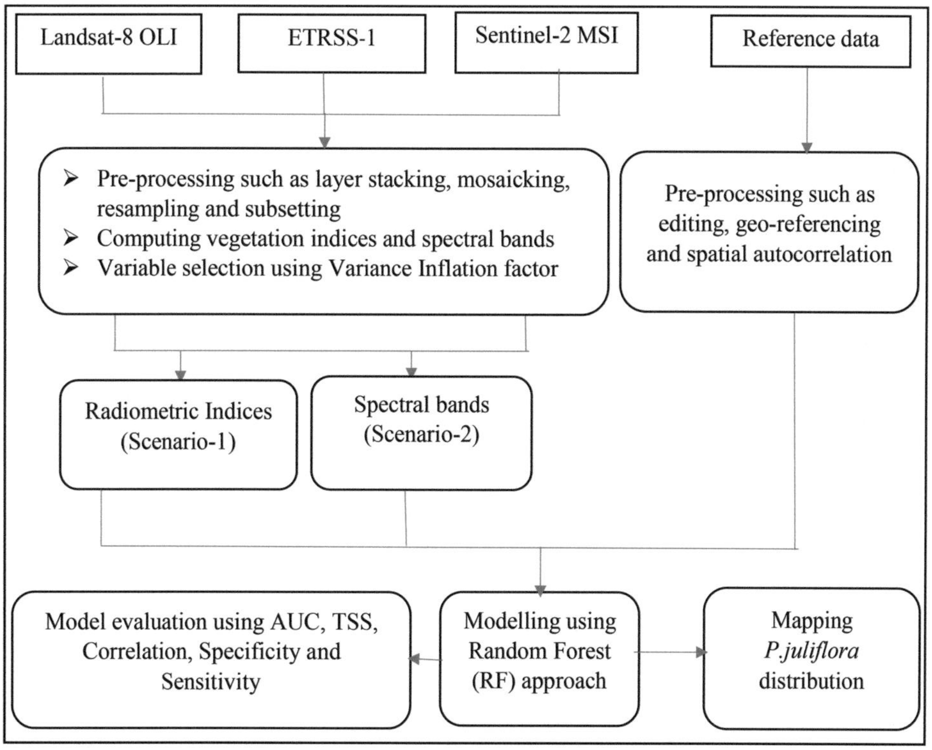

Figure 1.
Methodological flow chart describing the overall process of modelling P. juliflora distribution using sentinel-2B level 2A, ETRSS-1, and Landsat-8 OLI-derived VIs and spectral bands.

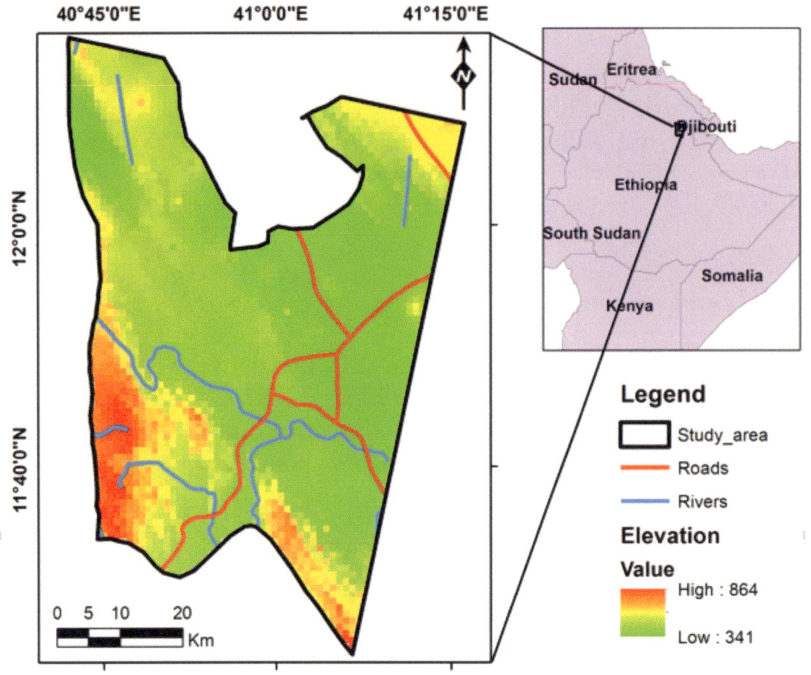

Figure 2.
Location of the study area (right) including roads, elevation, and rivers map of East Africa (right), including the location of the study area.

temperature at Aysaita and Dubti stations were reported as 160 mm & 214 mm and 28.1°C & 29°C, respectively. Bush-shrub and woodlands (30.5%), Barelands (25.4%), Grasslands (23.9%), and *P. juliflora* (14.5%) are the dominant land cover in the region [19]. Furthermore, *Acacia nilotica*, *Acacia robusta*, *Acacia Senegal,* and *Tamarix nilotica* are the dominant plant species along the Awash River basin [20].

The Awash River basin is the most widely used river for irrigation, agriculture, hydropower, and drinking water [21]. Pastoralism and agropastoralism are the dominant sources of livelihood [22, 23]. However, environmental challenges such as regular droughts, flooding, soil erosion and salinity, overgrazing, and the spread of invasive plant species, such as *P. juliflora,* are the major challenges affecting land use management in the area [24]. *P. juliflora* was purposely introduced in the early 1980s to support soil and water management and to combat drought and conservation in the dryland parts of the country [8, 25]. However, after a few years, it becomes invasive and poses a huge negative impact on agricultural and grazing lands, native grasses, and wetlands, which are the main sources of livelihood in the region [8, 26]. As a result, the Ethiopian government declared the species as a dangerous plant that needs to be controlled, managed, and if possible removed [27].

2.2 Data

2.2.1 Reference data

For modeling the current distribution of *P. juliflora*, georeferenced survey data on its occurrence and absence points were collected with the help of the Global

Positioning System (GPS) from January to February 2020. A stratified random sampling approach was used and 241 points were collected. Out of which, 30% were occurrences, while 70% were absence points. This share was chosen considering the previous distribution of *P. juliflora* in the area [18, 28]. *Moran's Index* was used to assess the spatial autocorrelation among observations. As a result, we found a *Moran's Index* of 0.28 and a Z-score of 2.45, suggesting no apparent spatial clustering among the points [29]. Furthermore, 70% and 30% of the collected data were used to train and test models, respectively [30].

2.2.2 Satellite data acquisition and processing

In this study, ETRSS-1 level-2B, Sentinel-2B level-2A, and Landsat-8 OLI level-2 were used. Detailed descriptions of satellite acquisition dates and their spatial characteristics are presented in **Tables 1** and **2**. Sentinel-2B level-2A dataset was acquired from European Space Agency (ESA) (https://scihub.copernicus.eu/dhus/#/home). It provides geometric and radiometrically corrected images. It also provides the bottom of the atmosphere (BOA) reflectance image converted from the level-1C top-of-atmosphere [31]. In addition, Landsat-8 OLI was acquired from the United States Geological Survey (USGS) (https://earthexplorer.usgs.gov/). Furthermore, Level-2B, ETRSS-1 acquired from the Ethiopian Space Science Technology Institute,

Satellite sensors	Date of acquisitions	Spatial resolutions in meter
ETRSS-1	08/03/2020	13.75
Landsat-8 OLI	29/02/2020 to 19/01/2020	30
Sentinel-2	19/01/2020 to 28/02/2020	10-20

Table 1.
Date of acquisitions and spatial resolutions of satellite sensors.

ETRSS-1 Landsat-8 OLI Sentinel-2					
Spectral bands	**WLR (μm)**	**Spectral bands**	**WLR (μm)**	**Spectral bands**	**WLR (μm)**
B1-B	0.45–0.52	B2-B	0.45–0.51	B2-B	0.458–0.523
B2-G	0.52–0.59	B3-G	0.53–0.59	B3-G	0.543–0.578
B3-R	0.63–0.69	B4-R	0.64–0.67	B4-R	0.650–0.680
—	—	—	—	B5-RE1	0.698–0.713
—	—	—	—	B6-RE2	0.733–0.748
—	—	—	—	B7-RE3	0.773–0.793
B4-NIR	0.77–0.89	—	—	B8-NIR	0.785–0.900
		B5-NIR	0.85–0.88	B8A-NNIR	0.855–0.875
		B6-SWIR1	1.57–1.65	B11-SWIR1	1.565–1.655
		B7-SWIR2	2.11–2.29	B12-SWIR2	2.1–2.28

Where WLR is wave length range, B is blue, G is green, R is red, NIR is near-infrared, SWIR is short wave infrared, RE is a red edge, and NNIR is narrow near-infrared.

Table 2.
Spectral bands and their wavelength range of satellite sensors used in the study.

Department of remote sensing development and research. Moreover, the radiometric calibration and atmospheric correction for both Landsat-8 OLI and ETRSS-1 were conducted using the FLAASH module in ENVI 5.3.

A total of four Sentinel-2B, three ETRSS-1, and two Landsat-8 OLI scenes were required to cover the study area. Bilinear interpolation was used to resample Sentinel-2B 20 M to 10 M pixel spacing. Pre-processing such as layer stacking, image mosaicking, and sub-setting for these scenes were processed using Sentinel Application (SNAP) 7.0, ENVI 5.3, and ArcGIS 10.8 software.

3. Methods

To evaluate the potential of ETRSS-1in modeling invasive *P. juliflora* distribution, two model scenarios were developed. The first scenario considers spectral bands derived from ETRSS-1 level-2B, Sentinel-2B Level-2A, and Landsat-8 level-2. Four bands of ETRSS-1, 10 bands of Sentinel-2B, and six bands of Landsat-8 were used. However, the three "atmospheric bands" (Band 1, 9, and 10) from Sentinel-2B level-2A and two "radar bands" (Bands 10 and 11), cirrus (band 9), and coastal aerosol (band 1) from Landsat-8 OLI level-2 were excluded from this study (**Table 2**).

The second scenario examines radiometric indices derived from the above three satellite datasets. Radiometric indices are a quantifiable measurement of a property that is produced by combining multiple spectral bands and which would not otherwise be apparent if only one band were used. Thirty commonly used radiometric indices were collected from the study of Ahmed et al. [6] and Ng et al. [7]. Twenty strongly collinear variables were excluded and 10 variables (**Table 3**) were maintained using the VIF correlation method with threshold values less than 0.9 [30, 32] in the R 4.0 software (R Development Core Team, 2021) "sdm" package [33]. This process was repeated for different sensors, and the common variables in all sensors (10 variables) with threshold values less than 0.9 were consistently used for different models.

The atmospherically resistant vegetation index algorithm (ARVI) describes its resistance to atmospheric effects and is accomplished by a self-correction process for the atmospheric effect on the red channel. This is done using the difference in the radiance between the blue and the red channels to correct the radiance in the red channel. The second brightness index (BI2) algorithm is representing the average brightness of a satellite image. The result looks like a panchromatic image with the same resolution as the original image. The color index (CI) is used for diachronic analyses and helps for a better understanding of the evolution of soil surfaces. The green normalized difference vegetation index (GNDVI) is used to identify different concentration rates of chlorophyll, which is highly correlated with nitrogen. The second modified soil adjusted vegetation index (MSAVI2) is an index created to replace NDVI when its accuracy is compromised by a lack of vegetation or chlorophyll in the plants. There is a great deal of bare soil between the seedlings, while they are in germination and leaf growth stages. The pigment specific simple ratio (PSSRa) investigates the potential of a range of spectral approaches for quantifying pigments at the scale of the whole plant canopy. The redness index (RI) is a type of soil radiometric index that is used to identify soil color variations. The transformed normalized difference vegetation index (TNDVI) algorithm indicates a relation between the amounts of green biomass that is found in a pixel. The transformed soil adjusted vegetation index algorithm (TSAVI) assumes that the soil line has an arbitrary slope and intercept, and

Radiometric indices	Description	Formula
ARVI	Atmospherically resistant vegetation index	ARVI = (IR_factor * near_IR – rb)/(IR_factor * near_IR + rb), where: rb = (red_factor * red) – gamma * (blue factor * blue – red_factor * red), with gamma = 1. Where IR = Infrared
BI2	The second brightness index	BI2 = sqrt(((red_factor * red * red_factor * red) + (green_factor * green * green_factor * green) + (IR_factor * near_IR * IR_factor * near_IR))/3) where sqrt = square root
CI	The Color Index	CI = (red_factor * red – green_factor * green)/(red_factor * red + green_factor * green)
GNDVI	Green normalized difference vegetation index	GNDVI = (IR_factor * near_IR – green_factor * green)/(IR_factor * near_IR + green_factor * green)
MSAVI2	The second modified soil adjusted vegetation index	MSAVI2 = (1/2) * (2 * IR_factor * near_IR + 1 – sqrt((2 * IR_factor * near_IR + 1) * (2 * IR_factor * near_IR + 1) – 8 * (IR_factor * near_IR - red_factor * red)))
PSSRa	The pigment specific simple ratio	PSSRa = (IR_factor * near_IR)/(red_factor * red)
RI	Redness index	RI = (red_factor * red * red_factor * red)/(green_factor * green * green_factor * green * green_factor * green)
TNDVI	Transformed normalized difference vegetation index	TNDVI = sqrt((IR_factor * near_IR – red_factor * red)/(IR_factor * near_IR + red_factor * red) + 0.5)
TSAVI	Transformed soil adjusted vegetation index	TSAVI = s * (IR_factor * near_IR – s * red_factor * red – a)/(s * IR_factor * near_IR + red_factor * red – a * s + X * (1 + s * s)) Where: a is the soil line intercept, – s is the soil line slope, X is the adjustment factor to minimize soil noise.
WDVI	The weighted difference vegetation index	WDVI = (IR_factor * near_IR – g * red_factor * red). Where: g is the slope of the soil line.

Table 3.
Radiometric indices derived from ETRSS-1, sentinel-2B level-2A, and Landsat-8 OLI used in this study. The different bands described in the formula for different sensors are presented in **Table 2**.

it makes use of these values to adjust the vegetation index. The *w*eighted *d*ifference *v*egetation *i*ndex (WDVI) is very sensitive to atmospheric variations. It provides a good adjustment for soil background when calculating the LAI of green vegetation.

3.1 Species distribution modeling and evaluation

Random forest (RF) approach was used to model the performance of all datasets in both scenarios. The robust RF approach received increasing interest owing to its high classification and prediction accuracy, and its ability to determine variable importance [34]. Moreover, its ability to provide out-of-bag errors helps for accurate predictions of invasive species [16]. The out-of-bag errors are also considered as internal validation of the dataset for each tree [35]. Several studies described the higher performance of the RF modeling approach compared to other models [6, 18, 36, 37].

Owing to this, the robust RF [38] is the most widely used machine learning algorithm in many remote sensing-based studies [7, 16, 34, 39–45]. Its ability to efficiently run large datasets and handle high data dimensionality also makes it preferable for remote sensing-based prediction [34]. Beside, the model can be efficiently used to select variables calculated from single or several sensors [46].

The 10-fold cross-validation approach was employed for the validation of model scenarios of all datasets. Considering the statistical capabilities and its advantage to control data processing and visualization R Version 4.0 software [47], the "sdm" package [33] was used. The cross-validation method provides a standardized framework and high computing performance for managing predictive models. It can also reduce the risk of overfitting models [36]. It divides the reference data into 10 sub-datasets at random, with each sub-dataset accounting for 10% of the total [33].

True skill statistics (TSS), area under the curve (AUC), sensitivity, specificity, and correlation were used to assess the relative performance of models. TSS shows how well the model can distinguish between matches and mismatches between observation and prediction, whereas AUC shows how well the model can identify presence and absence classes. In addition, sensitivity refers to a model's ability to predict occurrences where the species exist, whereas specificity refers to a model's ability to predict absences where a species does not exist. Furthermore, correlation describes the relationship between observation (field data) and prediction of the model. The threshold is a point that separates the presence (above the threshold) and absence (below the threshold) of a particular species [36].

Moreover, receiver operating characteristic (ROC) curves were produced for all datasets in both scenarios. Furthermore, the sensitivity-specificity sum maximization approach was employed to select the best threshold. This approach yields the most reliable prediction [48, 49]. Beside, for each dataset of a particular scenario, binary (invaded and uninvaded) maps were generated. Accordingly, pixels above and below the threshold represent the presence and absence of *P. juliflora* distribution, respectively. Beside, maps showing *P. juliflora* distribution and their percent coverage were computed in ArcGIS 10.8 software. This was computed considering the threshold value of the models (presence and absence values). The point above the threshold is considered as the occurrence of the species, while the point below the threshold is considered as absence of the species. Moreover, variable importance was derived using the "getVarImp" function in the R Version 4.0 software [47], the "sdm" package [33].

4. Results

4.1 Performances of satellite datasets

Figure 3 presented the relative performance of models derived from different satellite sensors. Accordingly, Sentinel-2-derived RIs (S2-RIs) showed higher performance with AUC = 0.97, and TSS = 0.89 followed by Landsat-8 OLI-derived RIs (L8-RIs) with AUC = 0.93 and TSS = 0.77 and ETRSS-1 RIs (ETRSS1-RIs) with AUC = 0.81 and TSS = 0.57. Furthermore, Sentinel-2 bands (S2-bands) showed higher performance with AUC = 0.97 and TSS = 0.93 followed by Landsat-8 OLI bands (L8-bands) with AUC = 0.93 and TSS = 0.79 and ETRSS1 bands (ETRSS1-bands) with AUC = 0.81 and TSS = 0.59.

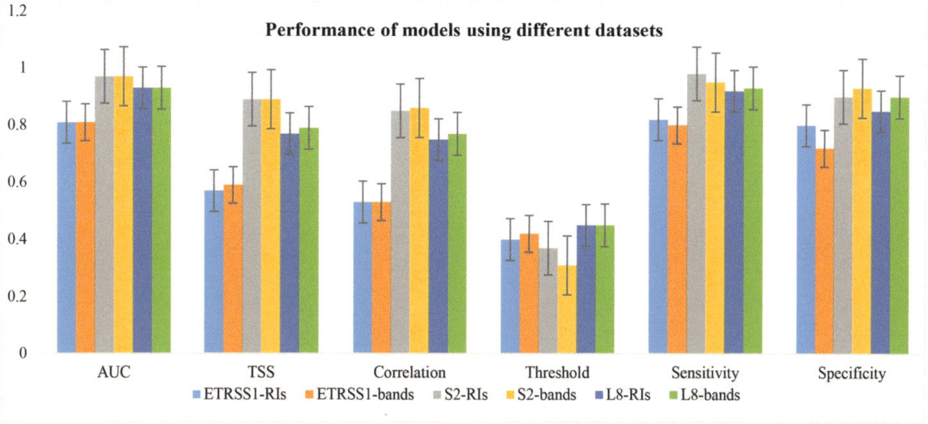

Figure 3.
The relative performance of models using ETRSS-1, Sentinel-2, and Landsat-8OLI-derived RIs and spectral bands. Sensitivity and specificity describe the rate of true-positive samples and false-positive samples (errors), respectively.

Furthermore, the receiver operator characteristics (ROC) curve (**Figure 4**) indicates the higher performance of Sentinel-2 (AUC > 0.97) followed by Landsat-8 OLI (AUC > 0.92) and ETRSS-1 (AUC > 0.8). ROC curve was computed using sensitivity (occurrence) and specificity (absence) values. The more the ROC curve approaches to the left the more the model becomes accurate and the more the ROC curve approaches to the diagonal line (right) the more the model needs improvement (see **Figure 3**).

4.2 Variable importance

Figure 5 described the relative influence of radiometric indices for all datasets. Accordingly, WDVI, GNDVI, and TSAVI from ETRSS1-RIs, MSAVI2, ARVI, and GNDVI from S2-RIs, ARVI, GNDVI, and BI2 from L8-RIs showed higher relative importance among the radiometric indices. In addition, blue, NIR, and green bands from ETRSS-1, blue, red, and NIR bands from Sentinel-2, and NIR, red, and green bands from Landsat-8 OLI showed higher relative influence compared to the other used spectral bands. In opposite, ARVI and CI from ETRSS1-RIs, BI2, RI, and CI from S2-RIs and CI and RI from L8-RIs showed lower relative performance indicating avoiding them and/or substituting with other variables will increase the accuracy of the model. In addition, NDVI, TNDVI, and PSSRa showed comparable performance for all datasets indicating using of them will better increase the accuracy of the model.

4.3 Distribution maps

Table 4 and **Figure 6** described the share of invasive *P. juliflora* distribution using radiometric indices and spectral bands for all datasets. Accordingly, the relative distribution of *P. juliflora* ranges from 4.3% of S2-bands to 10.8% of ETRSS1-RIs. In addition, the ETRSS1-RIs, ETRSS1-bands, and L8-RIs identified higher *P. juliflora* distribution compared to the other datasets.

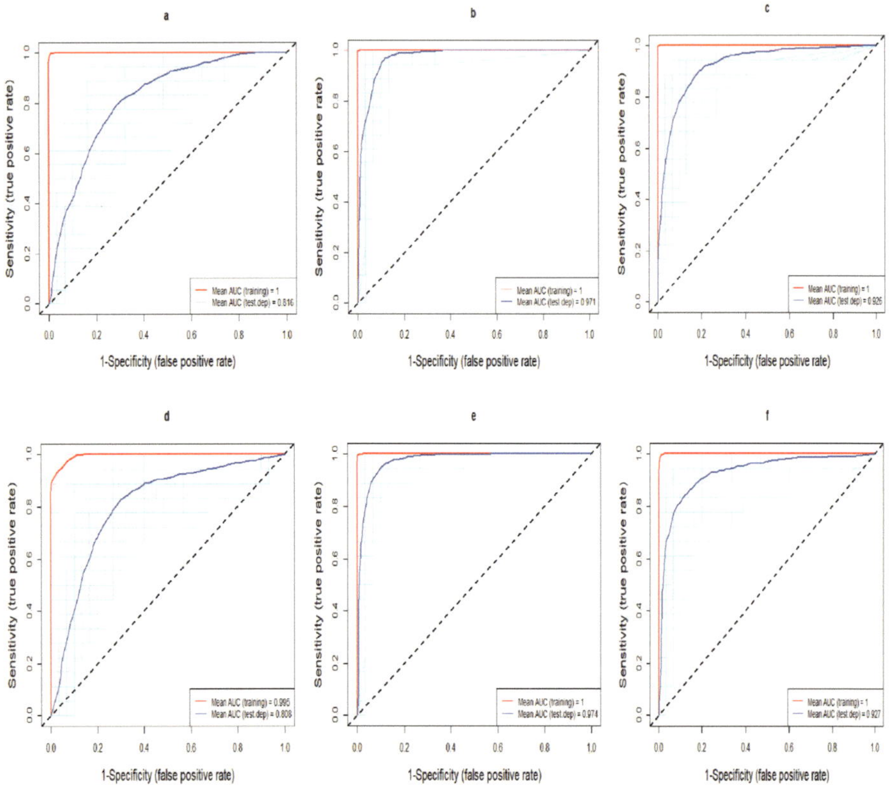

Figure 4.
AUC using training and test datasets for (a) ETRSS1-RIs, (b) S2-RIs, (c) L8-RIs, (d) ETRSS1-bands, (e) S2-bands, and (f) L8-bands.

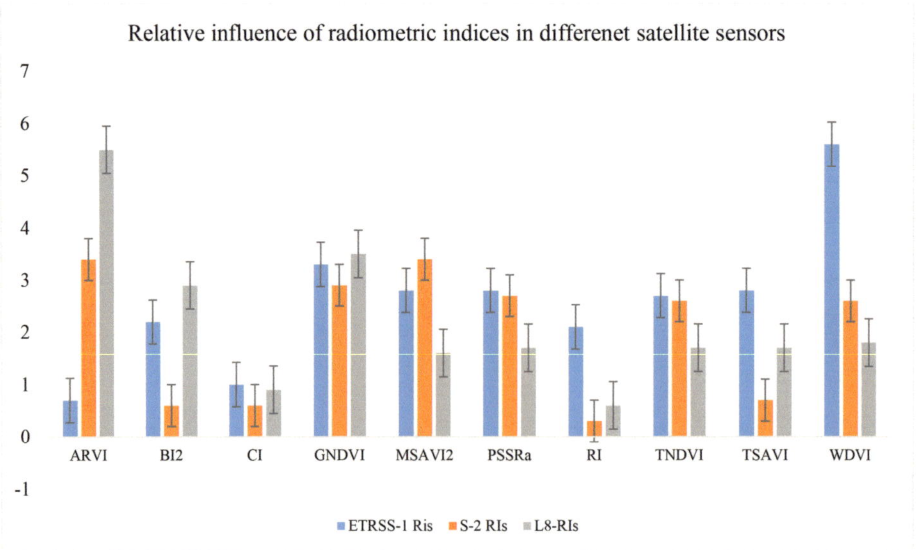

Figure 5.
The relative influence of models using ETRSS1-RIs, S2-RIs, and L8-RIs derived variables.

Datasets	Absence (%)	Presence (%)
ETRSS1-RIs	89.2	10.8
ETRSS1-bands	90.6	9.4
S2-RIs	94.1	5.9
S2-bands	95.7	4.3
L8-RIs	91.4	8.6
L8-bands	92.1	7.9

Table 4.
Proportion of P. juliflora invaded and uninvaded areas using vegetation indices and spectral bands of all datasets.

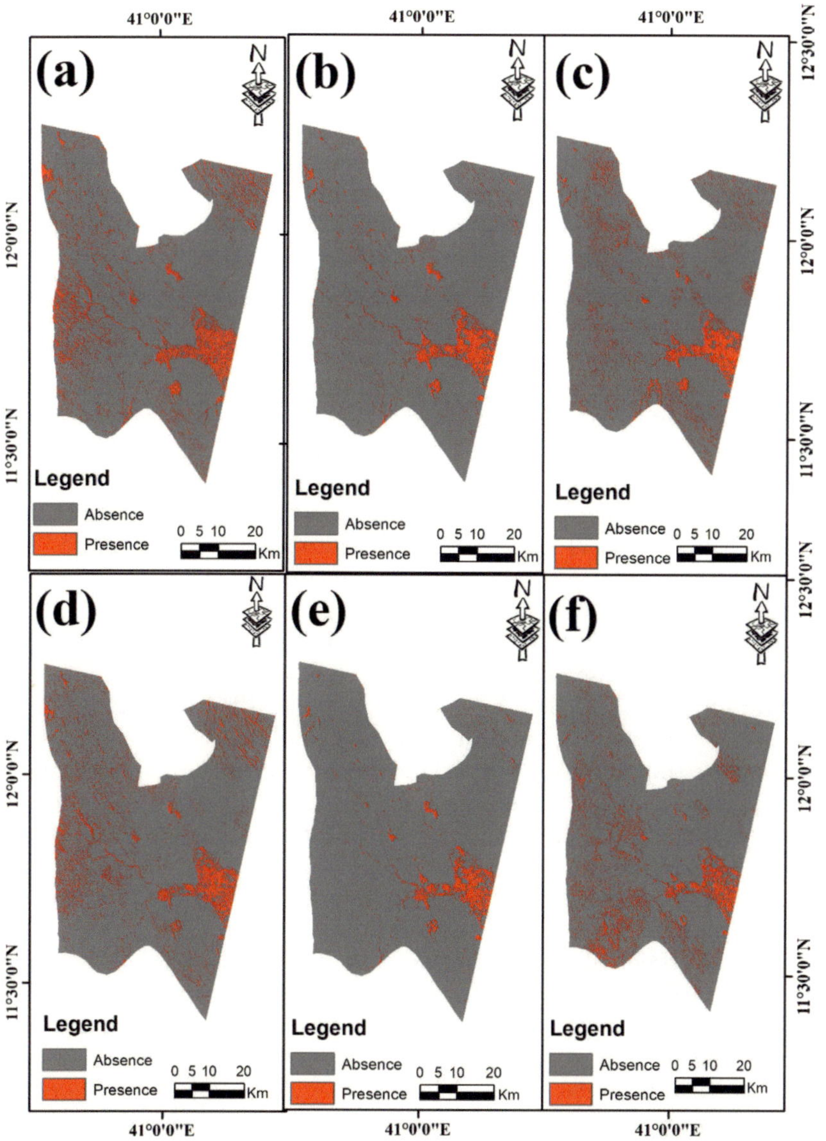

Figure 6.
P. juliflora distribution maps using different SDMs: (a) ETRSS1-RIs, (b) S2-RIs, (c) L8-RIs, (d) ETRSS1-bands, (e) S2-bands, and (f) L8-bands.

5. Discussion

Our study was intended to evaluate the potential of the newly launched satellite sensor, ETRSS-1, for detecting and mapping invasive *P. juliflora* distribution in the arid and semiarid regions of Ethiopia. To better evaluate its potential, a comparison was made with novel Landsat-8 OLI and Sentinel-2-derived variables. We employed both spectral bands and radiometric indices derived from each sensor. Accordingly, S2-RIs better performed with an AUC of 0.97 compared to Landsat-8 bands (AUC = 0.9) and ETRSS1-RIs (AUC = 0.83). Though its performance is lower than Sentinel-2 and Landsat-8 derived variables, ETRSS-1-derived variables showed average performance and can be sufficient for such kinds of studies. In both scenarios ETRSS-1-derived variables with AUC scores greater than 0.8 can be grouped in the category of "good" [50]. However, the capability of Sentinel-2 Landsat-8-based modeling was in the category of "excellent" as they scores greater than 0.9 of AUC.

Similar to our study, several studies described the higher performance of Sentinel-2-derived variables over Landsat-8-derived variables in the modeling of invasive species [14, 51]. Thamaga and Dube [14] tested the performance of Sentinel-2 and Landsat-8 derived RIs and spectral bands for mapping and modeling invasive *Waterhyathinth* and described the higher performance of Sentinel-2-derived variables. In addition, Dube et al. [51] evaluated the capability of Sentinel-2 and Landsat-8 spectral bands and vegetation indices for detecting and mapping invasive *Lantana Camara* and highlighted the higher accuracy of Sentinel-2 datasets. Owing to its higher spatial and spectral resolution, Sentinel-2 allows accurate identification and mapping of the fractional cover of invasive species [7, 14, 52]. Furthermore, the spectral features of four red-edge and three short wave infrared (SWIR) bands offer tremendous capacity for detecting and mapping species-level classification [16, 41, 53].

Our study also outlined the most important variables (spectral bands and radiometric indices) for detecting and mapping invasive *P. juliflora* distribution. Accordingly, WDVI, GNDVI, and TSAVI from ETRSS-1, MSAVI2, ARVI, and GNDVI from Sentinel-2 and ARVI, GNDVI, and BI2 from Landsat-8 showed higher relative contribution. The higher importance of vegetation indices, such as WDVI, NDVI, and TNDVI, are reported for mapping and modeling of invasive *P. juliflora* distribution to the study finding of Ahmed et al. [6] and Ng et al. [7]. In addition, B1 (blue) and B4 (NIR) from ETRSS-1, B2 (blue), B4 (red), and B8 (NIR) from Sentinel-2 and B5 (NIR) and B4 (red) from Landsat-8 provides higher relative contribution for modeling *P. juliflora* distribution. Our study is consistent with the study findings of Hoshino et al. [54], Mureriwa et al. [55], and Ng et al. [7]. A study by Ng et al. [7] described the higher importance of IR, blue, and green bands for detecting and mapping *P. juliflora* in Kenya. In addition, Mureriwa et al. [55] noted that *P. juliflora* is highly distinctive at NIR and narrow near-infrared bands. Moreover, *P. juliflora* showed high near-infrared and low reflectance at red bands [54].

Furthermore, comparable results between spectral bands and vegetation indices were found in all datasets indicating either spectral bands and/or radiometric indices can be applied for accurate identification and mapping of invasive species distribution. This finding is consistent with the study findings of Arogoundade et al. [9], Thamaga and Dube [14], and Dube et al. [51]. Thamaga and Dube [14] found comparable results between Landsat-8 VIs (accuracy = 65.53%) and spectral bands (accuracy = 63.34%) and Sentinel-2 VIs (accuracy = 73.31%) and spectral bands (accuracy = 73%) for mapping *water hyacinth*, South Africa. Our study also highlighted the higher impacts of spectral resolution on the accuracy of prediction. For example, the higher spatial resolution of ETRSS-1 (13.75 m) showed lower performance compared to Landsat-8 OLI (30 m) in both scenarios. This difference might occur partly due to their difference in spectral resolution.

6. Conclusion

In this study, we tested the potential of ETRSS-1 for detecting and mapping invasive *P. juliflora* distribution in the Afar regional state, Ethiopia. In addition, attempts were made to compare with Sentinel-2 and Landsat-8 OLI-derived variables. Radiometric indices (scenario-1) and spectral bands (scenario-2) were used for each dataset. Though it showed relatively lower performance, ETRSS-1 can be sufficient for mapping and identifying *P. juliflora* distribution. The lower performance of ETRSS-1 compared to Sentinel-2 and Landsat-8 might occur due to its coarse spectral resolution. Hence, increasing its spectral resolution with the help of different data fusion techniques might increase its accuracy.

Moreover, comparable results between spectral bands and radiometric indices were found in all datasets. This might indicate employing either spectral bands and/or radiometric indices can provide accurate identification and mapping of invasive species distribution. Accordingly, WDVI, MSAVI2, ARVI, and MNDVI from radiometric indices and B4 and B3 from spectral indices showed higher relative importance for detecting and mapping *P. juliflora* distribution. However, future works, such as web-based data accessibility and description of the ETRSS-1 dataset, are highly required.

Acknowledgements

This paper is part of the doctoral study entitled "Role of remote sensing in invasive species distribution modeling, the case of Prosopis in the lower Awash River basin, Ethiopia." We would like to thank the Ethiopian Space Science and Technology Institute (ESSTI) and Wollo University for allowing this doctoral study.

Funding

None.

Author contributions

All authors contributed to the study conception and design. All authors read and approved the final manuscript.

Competing interests

The authors declare that there is no conflict of interest.

Availability of data and materials

The datasets used and analyzed during the current study are available from the corresponding author.

Author details

Nurhussen Ahmed[1,2*] and Worku Zewdie[1]

1 Ethiopian Space Science and Technology Institute (ESSTI), Entoto Observatory and Research Center, Department of Remote Sensing, Addis Ababa, Ethiopia

2 Department of Geography and Environmental Studies, Wollo University, Dessie, Ethiopia

*Address all correspondence to: nurhussenahmed40@gmail.com

References

[1] Tatem AJ, Goetz SJ, Hay SI. Fifty years of earth observation satellites: Views from above have lead to countless advances on the ground in both scientific knowledge and daily life. American Scientist. 2008;**96**(5):390-398. DOI: 10.1511/2008.74.390.Fifty

[2] Woldai T. The status of earth observation (EO) & geo-information sciences in Africa – Trends and challenges. Geo-Spatial Information Science. 2020;**23**(1):107-123

[3] Chastain R, Housman I, Goldstein J, Finco M, Tenneson K. Empirical cross-sensor comparison of sentinel-2A and 2B MSI, Landsat-8 OLI, and Landsat-7 ETM + top of atmosphere spectral characteristics over the conterminous United States. Remote Sensing of Environment. 2019;**221**(2019):274-285. DOI: 10.1016/j.rse.2018.11.012

[4] Peerbhay K, Mutanga O, Ismail R. The identification and remote detection of alien invasive plants in commercial forests: An overview. South African Journal of Geomatics. 2016;**5**(1):49-67

[5] Royimani L, Mutanga O, Odindi J, Dube T. Advancements in satellite remote sensing for mapping and monitoring of alien invasive plant species (AIPs). Physics and Chemistry of the Earth. 2018;**112**:237-245. DOI: 10.1016/j.pce.2018.12.004

[6] Ahmed N, Atzberger C, Zewdie W. Species distribution modelling performance and its implication for Sentinel-2-based prediction of invasive Prosopis juliflora in lower Awash River basin, Ethiopia. Ecological Processes. 2021;**10**(18):1-16

[7] Ng W-T, Rima P, Einzmann K, Immitzer M, Atzberger C, Eckert S. Assessing the potential of sentinel-2 and pléiades data for the detection of prosopis and vachellia spp. In Kenya. Remote Sensing. 2017;**9**(74):1-29. DOI: 10.3390/rs9010074

[8] Wakie TT, Evangelista PH, Jarnevich CS, Laituri M. Mapping current and potential distribution of non-native Prosopis juliflorain in the Afar region of Ethiopia. PLoS One. 2014;**9**(11):e112854. DOI: 10.1371/journal.pone.0112854

[9] Arogoundade AM, Odindi J, Mutanga O. Modeling Parthenium hysterophorus invasion in KwaZulu-Natal province using remotely sensed data and environmental variables. Geocarto International. 2019;**35:13**:1-15. DOI: 10.1080/10106049.2019.1581268

[10] Rembold F, Leonardi U, Ng W-T, Gadain H, Meroni M, Atzberger C. Mapping areas invaded by Prosopis juliflora in Somaliland on Landsat 8 imagery. Remote Sensing for Agriculture, Ecosystems, and Hydrology XVII. 2015;**9637**(963723-1):295-306

[11] Shiferaw H, Schaffner U, Bewket W, Alamirew T, Zeleke G, Teketay D, et al. Modeling the current fractional cover of an invasive alien plant and drivers of its invasion in a dryland ecosystem. Scientific Reports. 2019a;**9**(1576):1-12

[12] Davis E, Wang C, Dow K, Davis E, Wang C. Comparing Sentinel-2 MSI and Landsat 8 OLI in soil salinity detection: A case study of agricultural lands in coastal North Carolina detection: A case study of agricultural lands in coastal north. International Journal of Remote Sensing. 2019;**00**(00):1-20. DOI: 10.1080/01431161.2019.1587205

[13] Costa S, Santos V, Melo D. Evaluation of Landsat 8 and sentinel - 2A data on the correlation between geological mapping and NDVI. Geoscience and Remote Sensing. 2017;**2017**:1-4

[14] Thamaga KH, Dube T. Testing two methods for mapping water hyacinth (Eichhornia crassipes) in the Greater Letaba river system, South Africa: Discrimination and mapping potential of the polar-orbiting Sentinel-2 MSI and Landsat 8 OLI sensors. International Journal of Remote Sensing. 2018;**39**(22):8041-8059

[15] Novelli A, Aguilar MA, Nemmaoui A, Aguilar FJ. Performance evaluation of object-based greenhouse detection from Sentinel-2 MSI and Landsat 8 OLI data: A case study from Almería (Spain). International Journal of Applied Earth Observations and Geoinformation. 2016;**52**:403-411. DOI: 10.1016/j.jag.2016.07.011

[16] Jensen T, Hass FS, Akbar MS, Petersen PH, Arsanjani JJ. Employing machine learning for detection of invasive species using sentinel-2 and aviris data: The case of kudzu in the United States. Sustainability. 2020;**12**(9):1-16. DOI: 10.3390/SU12093544

[17] Alvarez-taboada F, Paredes C, Julián-Pelaz J. Mapping of the invasive species Hakea sericea using unmanned aerial vehicle (UAV) and WorldView-2 imagery and an object-oriented approach. Remote Sensing. 2017;**9**(913):1-17

[18] Shiferaw H, Bewket W, Alamirew T, Zeleke G, Teketay D, Bekele K, et al. Implications of land use/land cover dynamics and Prosopis invasion on ecosystem service values in Afar region, Ethiopia. Science of the Total Environment. 2019b;**675**:354-366. DOI: 10.1016/j.scitotenv.2019.04.220

[19] Shiferaw H, Bewket W, Eckert S. Performances of machine learning algorithms for mapping fractional cover of an invasive plant species in a dryland ecosystem. Ecology and Evolution. 2019c;**9**(5):2562-2574. DOI: 10.1002/ece3.4919

[20] Tikssa M, Bekele T, Kelbessa E. Plant community distribution and variation along the awash river corridor in the main Ethiopian rift. African Journal of Ecology. 2009;**48**:21-28

[21] Mulugeta S, Fedler C, Ayana M. Analysis of long-term trends of annual and seasonal rainfall in the Awash River basin, Ethiopia. Water. 2019;**11**(1498):1-22

[22] Edossa DC, Babel MS, Gupta AD. Drought analysis in the Awash River basin, Ethiopia. Water Resources Management. 2010;**24**(7):1441-1460

[23] Tadese MT, Kumar L, Koech R, Zemadim B. Hydro-climatic variability: A characterization and trend study of the Awash River basin, Ethiopia. Hydrology. 2019;**6**(35):1-19

[24] ANRS. Afar National Regional State Rural Land Use and Administration Policy (Issue June). 2008. Available from: https://faolex.fao.org/docs/pdf/eth165071.pdf

[25] Tilahun M, Birner R, Ilukor J. Household-level preferences for mitigation of Prosopis juliflora invasion in the Afar region of Ethiopia: A contingent valuation. Journal of Environmental Planning and Management. 2017;**60**(2):282-308

[26] Ayanu Y, Jentsch A, Müller-Mahn D, Rettberg S, Romankiewicz C, Koellner T. Ecosystem engineer unleashed: Prosopis juliflora threatening ecosystem services? Regional Environmental Change.

2014;**15**(1):155-167. DOI: 10.1007/s10113-014-0616-x

[27] MoLF. Federal Democratic Republic of Ethiopia Ministry of Livestock and Fisheries National Strategy on *Prosopis Juliflora* Management. 2017

[28] Linders T, Bekele K, Schaffner U, Allan E, Alamirew T, Choge S, et al. The impact of invasive species on social-ecological systems: Relating supply and use of selected provisioning ecosystem services. Ecosystem Services. 2020;**41**(101055):1-14

[29] Abdulhafedh A. A novel hybrid method for measuring the spatial autocorrelation of vehicular crashes: Combining Moran's index and i statistic. Open Journal of Civil Engineering. 2017;**7**:208-221. DOI: 10.4236/ojce.2017.72013

[30] Engler R, Waser LT, Zimmermann NE, Schaub M, Berdos S, Ginzler C, et al. Combining ensemble modeling and remote sensing for mapping individual tree species at high spatial resolution. Forest Ecology and Management. 2013;**310**:64-73

[31] Szantoi Z, Strobl P. Copernicus Sentinel-2 Calibration and Validation. European Journal of RemoteSensing. 2019;**52**(1):253-255. DOI: 10.1080/22797254.2019.1582840

[32] Zimmermann NE, Edwards TC, Moisen GG, Frescino TS, Blackard JA. Remote sensing-based predictors improve distribution models of rare, early successional, and broadleaf tree species in Utah. Journal of Applied Ecology. 2007;**44**(5):1057-1067

[33] Naimi B, Araújo MB. Sdm: A reproducible and extensible R platform for species distribution modeling. Ecography. 2016;**39**:001-008. DOI: 10.1111/ecog.01881

[34] Rodriguez-galiano VF, Ghimire B, Rogan J, Chica-olmo M, Rigol-sanchez JP. An assessment of the effectiveness of a random forest classifier for land-cover classification. ISPRS Journal of Photogrammetry and Remote Sensing. 2012;**67**:93-104

[35] Boulesteix A, Janitza S, Kruppa J. Overview of random forest methodology and practical guidance with emphasis on computational biology and bioinformatics. WIREs Data Mining Knowl Discov. 2012;**2**:493-507. DOI: 10.1002/widm.1072

[36] Ng WT, de Oliveira C, Silva A, Rima P, Atzberger C, Immitzer M. Ensemble approach for potential habitat mapping of invasive Prosopis spp. in Turkana, Kenya. Ecology and Evolution. 2018;**8**(23):1-11. DOI: 10.1002/ece3.4649

[37] Schonlau M, Zou RY. The random forest algorithm for statistical learning. Stata Journal. 2020;**20**(1):3-29. DOI: 10.1177/1536867X20909688

[38] Breiman L. Random forests. Machine Learning. 2001;**45**:5-32. DOI: 10.1201/9780367816377-11

[39] Abdi AM. Land cover and land use classification performance of machine learning algorithms in a boreal landscape using Sentinel-2 data. GIScience and Remote Sensing. 2020;**57**(1):1-20. DOI: 10.1080/15481603.2019.1650447

[40] Immitzer M, Neuwirth M, Böck S, Brenner H, Vuolo F, Atzberger C. Optimal input features for tree species classification in Central Europe based on multi-temporal Sentinel-2 data. Remote Sensing. 2019;**11**(2599):1-23

[41] Immitzer M, Vuolo F, Atzberger C. First experience with Sentinel-2 data for crop and tree species classifications in Central Europe. Remote Sensing. 2016;**8**(166):1-27

[42] Kosicki JZ. Generalized additive models and random Forest approach as effective methods for predictive species density and functional species richness. Environmental and Ecological Statistics. 2020;**27**:273-292. DOI: 10.1007/s10651-020-00445-5

[43] Ma W, Feng Z, Cheng Z, Chen S, Wang F. Identifying forest fire driving factors and related impacts in China using random forest algorithm. Forests. 2020;**11**(507):1-26

[44] Pal M. International journal of remote random forest classifier for remote sensing classification. International Journal of Remote Sensing. 2007;**26**(1):217-222

[45] Sabat-Tomala A, Raczko E, Zagajewski B. Comparison of support vector machine and random forest algorithms for invasive and expansive species classification using airborne hyperspectral data. Remote Sensing. 2020;**12**:1-21. DOI: 10.3390/rs12030516

[46] Belgiu M, Dragut L. ISPRS Journal of Photogrammetry and Remote Sensing. 2016;**114**:24-31

[47] R Development Core Team. R A language and environment for statistical computing. R foundation for statistical computing, Vienna, Australia; 2020. Available from: https://www.r-project.org/index.html

[48] Jimenez-Valverde A, Lobo JM. Threshold criteria for conversion of probability of species presence to either-or presence-absence. Acta Oecologica. 2007;**31**:361-369

[49] Liu C, Berry PM, Dawson TP, Pearson RG. Selecting thresholds of occurrence in the prediction of species distributions. Ecography. 2005;**28**:385-393

[50] González-Ferreras AM, Barquín J, Peñas FJ. Integration of habitat models to predict fish distributions in several watersheds of northern Spain. Journal of Applied Ichthyology. 2016;**32**:204-216. DOI: 10.1111/jai.13024

[51] Dube T, Shoko C, Sibanda M, Madileng P, Maluleke XG, Mokwatedi VR, et al. Remote sensing of invasive Lantana camara (Verbenaceae) in semiarid savanna rangeland ecosystems of South Africa. Rangeland Ecology & Management. 2020;**73**(3):411-419. DOI: 10.1016/j.rama.2020.01.003

[52] Laurin GV, Puletti N, Hawthorne W, Liesenberg V, Corona P, Papale D, et al. Discrimination of tropical forest types, dominant species, and mapping of functional guilds by hyperspectral and simulated multispectral Sentinel-2 data. Remote Sensing of Environment. 2016;**176**:163-176. DOI: 10.1016/j.rse.2016.01.017

[53] Rajah P, Odindi J, Mutanga O, Kiala Z. The utility of Sentinel-2 vegetation indices (VIs) and Sentinel-1 synthetic aperture radar (SAR) for invasive alien species detection and mapping. Nature Conservation. 2019;**35**:41-61

[54] Hoshino B, Karamalla A, Abdelbasit MA, Manayeva K, Yoda K, Suliman M, et al. Evaluating the invasion strategic of Mesquite (Prosopis juliflora) in eastern Sudan using remotely sensed technique. Journal of Arid Land Studies ICAL. 2012;**4**:1-5

[55] Mureriwa N, Adam E, Sahu A, Tesfamichael S. Examining the spectral separability of Prosopis glandulosa from Co-existent species using field spectral measurement and guided regularized random forest. Remote Sensing. 2016;**8**(144):1-16

Chapter 6

Marine Geodesy for Hazard Investigation of EDOP Production Platform, Akwa Ibom State, Nigeria

Daniel Devote Basil, Lawrence Hart and Gbenga Ajayi

Abstract

These research focus on the application of marine geodesy in evaluating the potential risks that affect the positioning of any marine vessel while waiting on safe anchorage for further instruction in Mobil Oil Mining Lease (OML) 112, about 44-km offshore of Akwa Ibom state, Nigeria. Multibeam echo sounders, side-scan sonar and the shallow seismic recording system were used to investigate the seabed topography and shallow geology of EDOP PP vicinity. The results show that the area is stratigraphically subdivided into the upper soft silty clay unit, which grades into Clay and continues into Medium Sand to the penetration depth. The Side Scan Sonar imagery was studied carefully and interpreted accordingly. From the imagery, the area constitutes superficially soft Silty Clay sediment on the surface with some occasionally isolated strong sonar reflections. In the vicinity of the 500 m radius centered on EDOP PP, pipelines were observed to run out from the platform in the North-west, Eastern, and southern directions. Generally, these pipelines were observed to be exposed superficially along its length through to the coverage area.

Keywords: marine geodesy, shallow topography, subduction zone, bathymetry, hazard investigation

1. Introduction

The ocean and coastal water contain 28% percent of the global natural gas production and 37% percent of global oil production: the gold of the sea [1]. Many gas and oil fields have been discovered and exploited during the last century and the role of methane hydrate, which can be found on the floor of almost all oceans, for industrial use has not even been exploited. The increasing demand for all kinds of sea food has also created a large and still growing fishing industry which will satisfy their clams in national and international waters [2]. More so, the current transition to cleaner and greener energy have yet presented another potential for the sea: Wind energy. The fastest way to create substantial amount of this green energy is to build offshore wind farms. The seas have space enough, the wind yield is higher than on shore and the visible impact, which is so a difficult hurdle to overcome on shore, is not an issue in the sea [3].

These coastal water and ocean surfaces have been the driving force of the climatic change of the planet earth. It is also the place where life started to develop and still the largest biosphere on earth [4, 5]. Ocean and coastal waters are not only important from the ecological point of views, but also from the economic aspect. It controls the world commerce as 90% percent of international exchange of data, information, goods, and service around the world are transported through either fiber optical cables buried in the sea bottom or large going ships that are getting bigger and bigger every day [6, 7].

Despite the tremendous potentials and natural resources of the Ocean surfaces and coastal waters which makes up 70% of the earth surface, unfortunately, the ocean surface and coastal water have not been adequately mapped. According to National Aeronautics and Space Administration (NASA), there are complete maps of the surface of the moon or mars than the ocean floor of the earth [8]. The mapping of the earth surfaces is the one of the major applications of geodesy [9]. Hence, it suffices to say that marine geodesy is the application of geodetic techniques in mapping of ocean and coastal environment, sea surface topography and marine geoid. Marine geodesy involves scientific investigations related to geology, geophysics, and glaciology [10].

Mapping topography of the ocean bottom has been very elusive for scientist and marine geodesist to achieve. The inhospitable underwater pressure and sheer expanse of the marine environment (70% of the earth surface) being the principal cause [11]. However, acoustic remote sensing techniques (sonar and seismic reflection profiling) have evolved within the last century allowing researchers the ability to effectively interpret and map large portion of benthonic environments [12].

Seafloor mapping is the first step in making a census of the geohazard-bearing features present within the seafloor [11]. It often provides the only tool for a comprehensive, although non-specific, seafloor geohazard assessment over large areas that are scarcely ground-truthed by acoustic prospection and seafloor sampling. Many hydrographic survey and visualization surveys of seafloors are primarily conducted acoustically using the single-beam echosounder, multibeam echosounder and Side Scan Sonar (Sound navigation and ranging) (SSS), which utilizes acoustic signals to determine the seabed topography.

Single-beam and Multibeam Echosounder S/MBES are acoustic sensors used in the mapping of underwater topography. These acoustic sensors transmit sound wave directly downward from the transducer that travels through the water column, and is reflected back to the transducers. The two-way time of travel of the acoustic signal is used to determine underwater depth [13]. Single-beam echo sounder reflects single point beneath the sonar, while multibeam echo sounder illuminates a swath elongated along across the bottom and perpendicular to the direction of the sailing [14]. To accurately determine and measure the direction which the acoustic signal was received at the transducer head, the position, heading, motion of the vessel, and variation in the speed of sound through the water column must be accurately measured, and time stamped in real time [15]. The operational principle of SSS is based on time series recording of the travel time and amplitude of acoustic signals from the SSS through the water column. These signals are being reflected in the direction of the SSS from the seabed and other benthic obstruction, the back-scatter effect [16, 17]. The strength of the return echo reflection is continuously recorded, creating a picture of the ocean bottom. SSS relies on the physics that the strength of the return signal is directly proportional to the reflectivity of the seabed material called spectral signature. In basic terms, strong reflectors create strong echoes, while weak reflectors create weaker echoes. Knowing these characteristics, researchers, and marine geodesist such as [17–19] have been able to use the strength of acoustic returns from side scan sonar to examine the composition of the sea floor.

However, the characterization of geohazard features on a morphological basis alone is limited, and more detailed investigations are needed to define the character and state of activity of potentially hazardous features. Such investigations include the use of deep-tow or autonomous platforms designed to acquire high-resolution data at depth as well as *in situ* measurements, both being very expensive activities not applicable over large areas. Thus, seafloor mapping is often not only the first and the main but also the only tool for a comprehensive seafloor geohazard assessment [20].

Several regions of the world's oceans and coastal areas lack detailed seafloor mapping data, leading to significant gaps in our understanding of the underwater terrain. This lack of coverage hampers hazard assessment efforts, making it difficult to predict and mitigate potential threats. Oil Mining Lease (OML) 112 is an offshore oil drilling facility, which is operated by the Mobil oil producing company Ltd. Due to the ongoing exploration and exploitation activities, several ships and tankers usually navigate and anchor within the vicinity of the production facility. The need for regular seafloor mapping of the production site is critical not just for navigational purpose, but also to ensure safe anchoring without damaging oil delivering pipelines buried under the seabed. This study focuses on the application of seafloor mapping in evaluating potential risks that affect positioning and safe anchorage of marine vessel in Mobil Oil Mining Lease (OML) 70, about 44-km offshore of Akwa Ibom state, Nigeria. Its fundamental objective is the identification of potential hazard on, and shallow geology of the benthonic area. It involves the complete sonification of OML 112 offshore Mobil oil platform areas using acoustic sensors.

Sonar and Seismic profiling surveys have been used to map the benthic regions since the nineteenth century with marine reflection surveys coming to prominence in the 1950s when they were used in acoustic bathymetry, seabed identification using multibeam and side scan sonar instrument [21]. Side scan sonar investigations and marine seismic identification of seabed object were conducted in Punggur waters, Indonesia. C-Max (CM2) instrument was deployed connected with a 50-meter cable the 7–16-m altitudes above the seabed. The results obtained from side scan sonar recording are with high-resolution seabed imagery. The side scan sonar imagery investigation shows 4 objects detected at 187.8, 137.1, 70.9, and 23.7 m. The increased knowledge about pockmark features have resulted from this survey, it was mainly achieved by the side scan sonar which was towed at an optimum altitude (10–26 m) above the bottom, regardless of the (actual) water depth [22].

1.1 Study area

The study area is in OML 112 (AFREN Field) which is in Ikot Abasi Local Government Area of Akwa Ibom State. It is in the South – Western part of the state on Nigerian territorial waters in the Gulf of Guinea, Nigeria. OML 112 is a located within latitude 3° 50 3.74" N and 3° 38 42.82" N, and latitude 7° 47 02.85″ E and 8° 02 36.99″ E, belonging to the Exxon Mobil oil company, Nigeria (**Figure 1**).

Figure 1.
Study area.

S/N	Service	Equipment
1	Surface positioning	C-Nav 3050 DGPS
2	Surface heading primary	Meridian Surveyor Gyrocompass
3	Navigation software	EIVA-Navipac
4	Motion Reference Unit	Teledyne DMS 05
5	Multi beam echo sounder	Kongsberg GeoAcoustic GeoSwath Plus
6	Single beam echo sounder	ELAC Hydrostar 4300 Dual Frequency
7	Side scan sonar	EdgeTech & CMAX Dual Frequency
8	Sub-bottom profiler	EdgeTech X-Star SBP System
9	Magnetometer	Geometrics G882 Marine Magnetometer
10	SVP sensor	Valeport MiniSVP
11	MBES Acquisition	GS+ & Hypack Hysweep
12	SSS Acquisition	EdgeTech Discover & Maxview
13	SBP Acquisition	EdgeTech Discover
14	Position Calibration and offsets	Leica TCR 705 total station
15	Data processing and cartography	SonarWiz, Hypack, GS+, MagPick, MicroStation, AutoCAD

Table 1.
List of equipment.

2. Material and methods

The study relies on primary data acquires from the complete sonification of the benthonic area. It entails the deployment of the following equipment given in **Table 1**.

2.1 Data acquisition

The data acquisition involves the setup, calibration of various sensor onboard the survey vessel (MV C- RACER) as show in **Figure 2**, and complete sonification of the study area. The setup involves measurement of sensors offset with respect to the vessel

Figure 2.
The set-up configuration of the various sensors.

SN	Equipment	DX (meters)	DY (meters)	DZ (meters)
1	CRP	0	0	0
2	DGPS Antenna (Secondary)	1.286	−16.141	6.94
3	Motion Reference Unit	0.657	−13.229	3.23
4	DGPS Antenna (Primary)	1.119	−10.664	6.94
5	Magnetometer (Tow Point)	−6.71	−36.214	—
6	Sub bottom Profiler (Tow Point)	0.42	−36.174	—
7	Side Scan Sonar (Tow Point)	6.716	−36.139	—
8	SBES Transducer	8.965	−20.733	−3.78
9	MBES Transducer	8.965	−20.033	−3.78

Table 2.
Offset measurement of the survey vessel.

reference systems (VFR). The vessel Central Reference Point (CRP) was taken as the origin of measurements for determining the horizontal positions of installed sensors, while the deck was taken as the reference for the vertical position of all installed sensors on the survey vessel. The offset positions as determined are listed in **Table 2**.

2.2 Sensors calibration

The calibration involves rub test, patch test, and chirp test for the side scan sonar, multi-beam echo sounder, and sub bottom profiler respectively.

2.2.1 Rub test

The result of rub test (rubbing the sonar sensors of the side scan sonar by hand) on the two channels to ascertain its functionality is as shown in **Figure 3**. This confirms the sensitivity of sonar sensors. The Edgetech dual frequency SSS has a two (2) channel side scan display, image correction capability, selectable color palettes, a tow cable of 1500 m, towing speed of 3–5 knots, frequency of 100 kHz – 105 ± 10 kHz; 500* kHz – 390 ± 20 kHz, and a scan range of 25 to 600 m (100 kHz – 500 m swath,

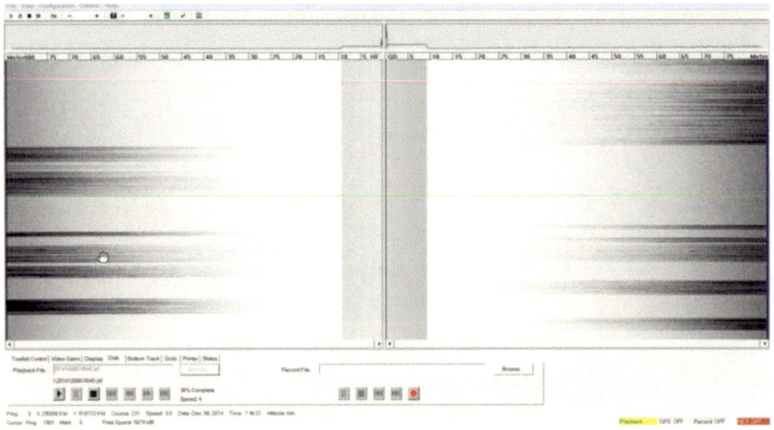

Figure 3.
Side scan sonar rub test.

500* kHz – 200 m swath). The CMAX dual frequency SSS has a range from 100 m to 500 m at 100 kHz, 25 m – 150 m at 325 kHz and 12.5 m – 37.5 m at 780 kHz. It has an operating depth from 0 to 2000 m, operating speed of 1–8 knots and a maximum towing speed of 12 knots, a water temperature sensor, and a tow cable of about 1000 m [16].

2.2.2 Patch test

The Multi-Beam Patch test involves the determination of the mounting angle (Roll, Pitch, and Heading) of the multi-beam sonar head, with the motion sensor and gyroscope, and the sychronisation of time (latency) between the sensors [15]. It is critical for the conversion of the slang range of the swath sonar to it equivalent depth, and correction of the dynamic motion of ocean surfaces. The determination of the calibration value was done using the classical method, which requires a pre-knowledge of the terrain characteristic of the seabed. Roll biases estimation requires surveying a flat seabed on opposite direction, for the pitch biases, a sloping seabed is required while the yaw (heading or azimuth) biases requires surveying a flat seabed terrain from two parallel and overlapping survey tracks heading in the same direction [23]. The result of the patch test is as given in **Figure 4**.

2.2.3 Magnetometer field test

The test of the magnetometer involves the introduction of a magnetic fluctuation within the magnetic field of the magnetometer. The tow fish was placed on the vessel deck away from any magnetic material while the system was started. Fluctuations in the magnetic field around the tow fish were noticed as soon as the magnetic material got close to the tow fish, increasing in signal strength as the material gets closer to the tow fish, and disappearing as soon as the material was further away as shown in **Figure 5**.

The Geometrics 882 magnetometer has an operating range from 20,000 to 100,000 nT, absolute accuracy of <2 nT throughout range, operates within

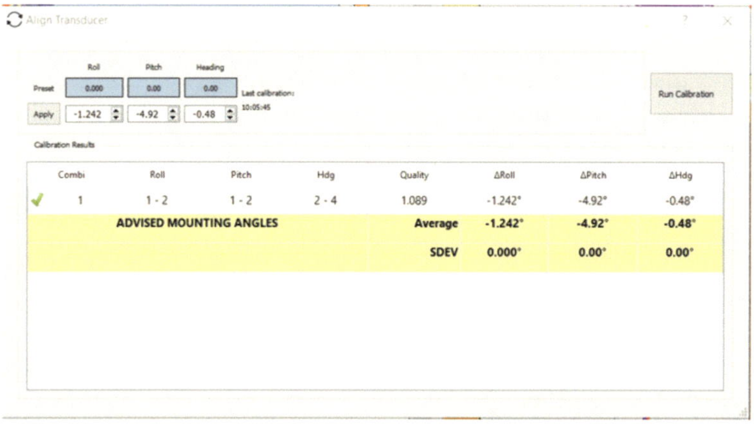

Figure 4.
The result of the multi beam patch test.

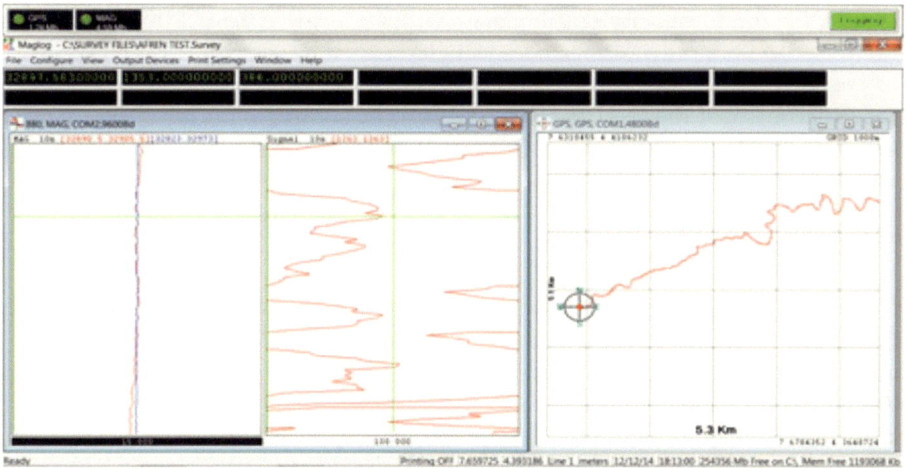

Figure 5.
Magnetometer field test.

temperature of −30°F to +122°F (−35°C to +50°C), maximum operating depth of 9000 m and a tow cable of 500 m.

2.2.4 Sub bottom profiler chirp test

The sub bottom Profiler (SBP) was tested to confirm that it was in good working condition by carrying out the 'chirp test'. This was done by setting up the system with the sub-bottom towfish placed on the deck of the vessel. The system is turned on and acquisition started, while the towfish is monitored. The towfish is expected to emit a 'chirp' sound to confirm that it is in good working condition. The Edgetech sub bottom profiler has a vertical resolution of 4–8 cm, frequency range between 4 and 24 kHz, a typical penetration depth of 2 m in coarse calcareous sand and 40 m in clay, and a maximum operating depth of 300 m.

3. Sonification of the study area

Data acquisition was done simultaneously on all sensors and the output from each sensor was logged digitally from the various systems/software. A log of all events, time and quality of data acquired was also kept.

3.1 Bathymetry (Single Beam Echo Sounder & Multi-Beam Echo Sounder)

The Kongsberg-GeoAcoustics GeoSwath+ MBES and ELAC Hydrostar 4300 dual frequency (30 kHz/200 kHz) SBES were used to acquire bathymetry data during the survey. The transducers were installed rigidly to their brackets and side mounted on the survey vessel. The transducer shoe is sufficiently deep and well positioned not to experience turbulence and aeration from the vessel during data acquisition.

Hydroacoustic method of depth determination is based on the propagation of sound wave through the water column. However, the speed of sound wave is refracted due to temperature, pressure and conductivity [6, 15]. Hence, the speed of sound wave needs to be accounted for accurate representation of the seabed. For this study, the sound velocity profiles taken with a measured using Mini-SVP, which was imputed in both the multibeam and single-beam data acquisition and processing software.

Quality control of the acquired data was done by comparing the difference between the single-beam and the multibeam bathymetric data, which was observed to be within ±20 cm.

3.2 Side scan sonar

The EdgeTech 4200 Dual frequency (100/500 kHz) and C-MAX Dual Frequency (325/780 kHz) digital Side Scan Sonar Systems were deployed and tuned between 75 m and 100 m slant range to scan the seabed for debris and seabed features. However, based on the field situation (depth variation) the range was tuned for better data quality ensuring a theoretical 75% /100% overlap on 100 m line spacing. Cable out was noted at the start of each line and whenever there was alterations. The output from the SSS essentially provided detailed representation of the sea bottom features and characteristics of the seabed.

3.3 Sub-bottom profiler

An EdgeTech X-Star Sub-bottom profiler, a high-resolution and greater penetration subsurface profiler with SB-216S tow-fish (2–16 kHz) was deployed to investigate the seabed and the depth to reflecting discontinuities in the near sub-surface. This was used for the shallow seismic data acquisition during the survey.

3.4 Magnetometer

The Geometrics G882 marine magnetometer was used for data acquisition. The data was displayed as profiles on the acquisition system. The magnetometer was towed at 150 m astern of the survey vessel and operated on the 32,000–33,200 Nano Tesla (nT) range, which is the typical range of magnetic field of this Latitude/Longitude Positions. The Geometrics MagLog software was used to log the magnetometer data. The software was interfaced to the DGPS directly. Different sub-windows setups during acquisition were navigation, signal strength, magnetic field data and depth showing data in numeric and graphical format. The simultaneous display of this information were used to perform the online QC Checks of the data quality.

The presence of pipelines and other metallic objects within the pipeline survey area were detected with the magnetometer. These were observed as deviation from the background value. Their degrees of exposure were seen as a result of their strong signal strength.

3.5 Data processing

The data acquired from individual sensors were processed using quality integrated navigation software (Qinsy) to produce the digital bottom model and the subsea geomorphological images, which requires geophysical interpretation. Digitally recorded depth data (SBES & MBES) were referenced to Lowest Astronomical Tide (LAT) in meters by correcting for tidal levels using tidal prediction tables for OPOBO River Entrance (a standard port). The height of water, referenced to LAT, was subtracted from the depth measurement to refer it to datum.

The side scan sonar imageries logged digitally were processed to produce a mosaic that was plotted in plain view. This plot was used to characterize the sea bottom. Bottom characterization was accomplished by interpreting the side-scan sonar records in conjunction with sounding data, sub-bottom data, and visual observations. The high-resolution sonar sub-bottom data was processed and analyzed to show the detailed sub-bottom profile of the surveyed area. The MagPick Software package was used to process the magnetometer data. Internal functions available in MagPick were used to perform the Projection Transformation. All the magnetic anomalies of the survey area were picked manually as the Manual Target Picking option offers more flexibilities of the surveyed area.

4. Result presentation and analysis

The data processed from various sensors was analyzed to give an overview of the seabed characteristics along the surveyed corridor.

4.1 Bathymetric result

The bathymetry study of the seabed was done using data from the SBES and MBES and presented in **Table 3** and **Figure 6**. The result of the bathymetric survey

KP (Km)	Eastings (m)	Northings (m)	Water Depth (m)
60	595725.91	44984.48	11.7
57.5	594012.764	45089.891	11.2
55	591512.767	45085.942	10
52.5	589016.15	44977.328	9.4
50	586520.928	44822.843	8.9
47.5	584025.705	44668.358	8.2

Table 3.
Water depth within the study area.

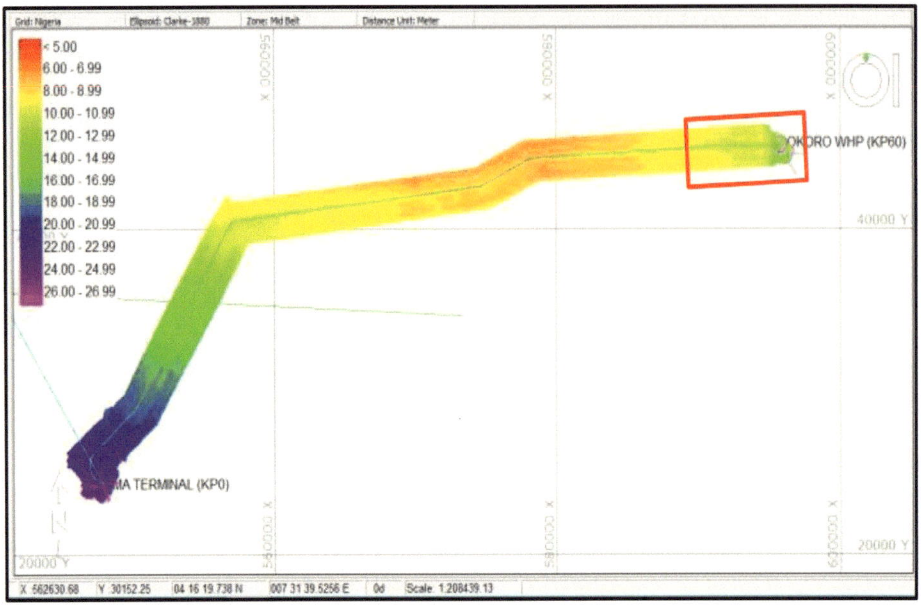

Figure 6.
MBES bathymetry DTM of KP0 to KP60 with pipeline route and structures underlay (KP50 to KP60 in red rectangle).

show that the study area has an average depth of 10 m with a maximum depth of 12 m and a minimum depth of 8 m.

4.2 Side scan sonar interpretation

The analysis of the Side Scan Sonar imagery presented is presented in **Table 4** below. As can be seen in **Figures 7** and **8**, the surveyed corridor exhibits a smooth texture, with a rough surface furrowed at some points. There were physical scars which were suspected to be anchor drags, as well as some linear features suspected to be a result of fishing activities in the area. Various sonar contacts were picked as well as seabed sediments.

4.3 Sub-bottom profile interpretation

The shallow geology of the surveyed corridor was analyzed from the Sub-Bottom Profiler data as shown in **Figure 9**. The seismic profiles of the surveyed area suggest the presence of a weak seismo-stratigraphic layer just below the seabed. It varies in thickness from about 4 m at the OKORO Well Head Platform end. No contact was made within this section of the study.

4.4 Magnetometer contact

Within the study area, only two (2) magnetic anomalies were observed along the surveyed corridor, and their details are as presented in **Table 5** and **Figure 10**. This shows the presence of magnetic materials below the seabed, which was suspected to be ships anchors.

Side Scan Sonar Features (KP60 – KP50)						
Features ID	**East (m)**	**North (m)**	**Length (m)**	**Width (m)**	**Height (m)**	**Description**
SSS 01	592329.962	45526.36	4.8	2.7	—	Debris
SSS 02	586157.582	45227.859	3.2	1.1	—	Debris
SSS 03	585572.604	45279.128	31.6	10.6	—	Seabed disturbance with Debris
SSS 04	589012.508	45482.766	45.3	9.8	—	Strong sonar reflection
SSS 05	585411.79	45013.966	Continuous	53	—	Edge of thick mud patch with erosional surface
SSS 06	592991.73	45320.295	9.8	7.3	—	Debris
SSS 07	585386.8	44822.551	7.9	4.1	—	Debris
SSS 08	589115.94	44589.034	8.4	1.2	0.5	Debris
SSS 09	584311.91	44748.303	Continuous	37	—	Set of linear scars
SSS 10	592922.73	44524.206	5	0.2	—	Stretch of linear object
SSS 11	592591.08	44501.67	34.8	14.3	—	Object on seabed
SSS 12	584189.26	44013.989	16.3	2.4	—	Object on seabed
SSS 13	585685.7	44102.544	8.1	4.5	0.2	Object on seabed
SSS 14	588448.79	45701.67	19.5	6.3	—	Group of Debris
SSS 15	585457.96	45486.53	150	10.7	—	Stretch of littered isolated debris
SSS 16	594598.91	45410.745	10.3	0.5	—	Linear object on seabed
SSS 17	595636.23	45294.642	5.7	2.1	—	Object on seabed
SSS 18	595401.09	45226.313	20.2	19.8	—	Spud can
SSS 19	595443.05	45202.691	19.2	18.7	—	Spud can
SSS 20	595434.95	45249.314	20.5	19.3	—	Spud can
SSS 21	595704.76	45192.873	20.5	19.3	—	Spud can
SSS 22	595742.83	45168.701	19.2	19.7	—	Spud can
SSS 23	595744.35	45212.552	21.5	1.7	—	Spud can
SSS 24	595770.96	45186.548	10.4	9.6	—	Platform
SSS 25	595702.49	44942.109	Continuous	—	—	Observed flexible pipes from the platform clamped down with concrete mattresses
SSS 26	595757.16	45051.505	Continuous	—	—	Observed flexible pipes from the platform clamped down with concrete mattresses

Side Scan Sonar Features (KP60 – KP50)						
Features ID	**East (m)**	**North (m)**	**Length (m)**	**Width (m)**	**Height (m)**	**Description**
SSS 27	595800.5	45127.561	Continuous	—	—	Observed flexible pipes from the platform clamped down with concrete mattresses

Table 4.
Analysis of the side scan sonar feature (KP60 – KP 50).

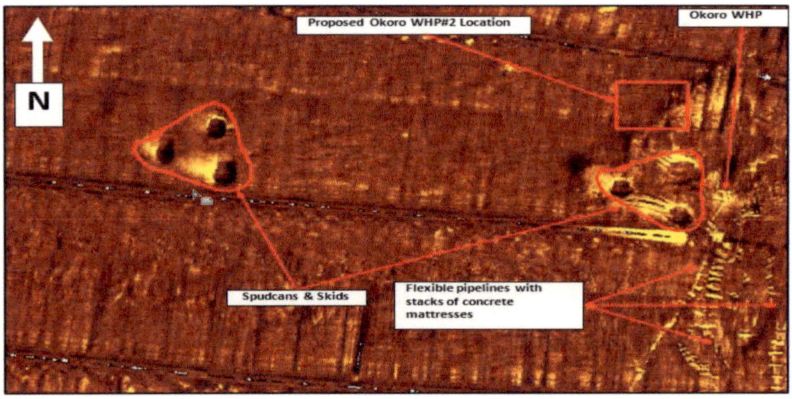

Figure 7.
MBES bathymetry DTM of KP0 to KP60 with pipeline route and structures underlay (KP50 to KP60 in red rectangle).

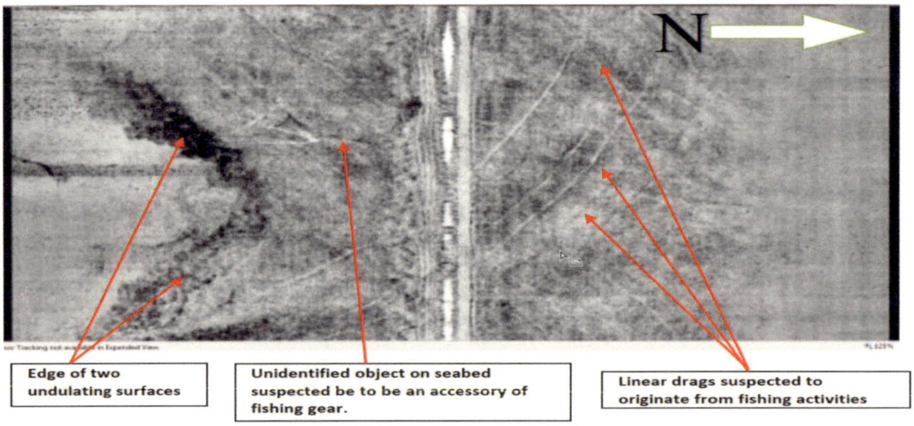

Figure 8.
Edgetech mosaic showing activities on the seabed.

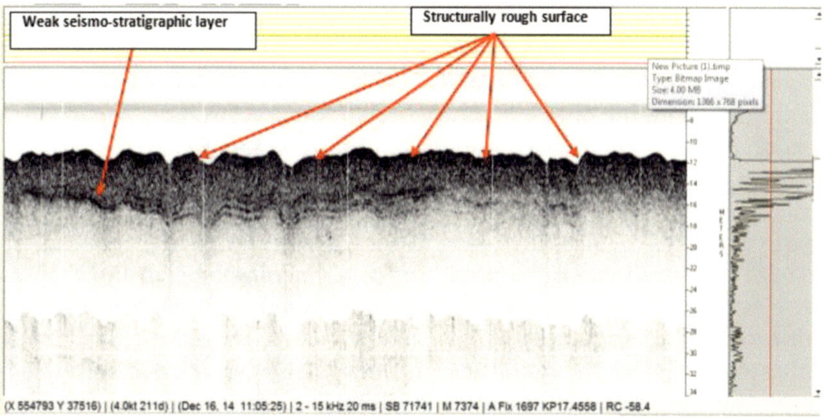

Figure 9.
Sub-bottom profile showing surface structural imperfection within the surveyed corridor.

Magnetometer Contacts (KP60 – KP50)				
Anomaly ID	**East (m)**	**North (m)**	**Amplitude (nT)**	**Type of Anomaly**
MAG 01	589256.534	45078.926	12	Negative
MAG 02	588876.059	45149.887	3	Negative

Table 5.
Magnetic contact list.

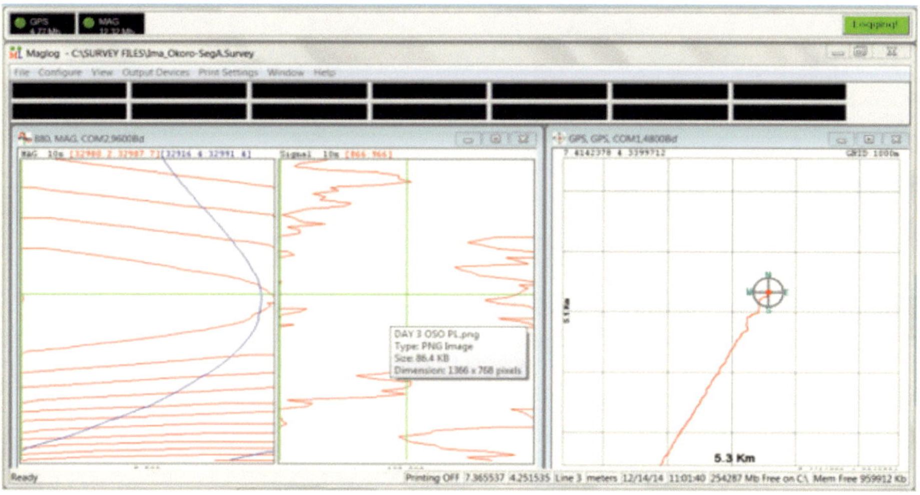

Figure 10.
Magnetic data distortion above Oso gas pipeline crossing.

5. Conclusion

This study shows the application of marine geodesy in evaluating potential risks that affect the positioning of any marine vessel in Akwa Ibom state, Nigeria using hydroacoustic sensors. Marine geodesy is a multidisciplinary field that provides essential data and tools for hazard investigation and mitigation in coastal and underwater environments. The increasing development of offshore and coastal facilities and growth of marine services involving mineral exploration and exploitation coupled with advances in instrumentation and methodology shows the potential of this emerging field. Marine geodesy has a very wide scope of work with a large variety of activities. It has become a nearly independent field of research with relations to many other areas of geodesy and influences from geology, geophysics, glaciology, and navigation. Marine Geodesy is not just interesting from the scientific point of view, but marine geodesy is also a lucrative business area for geodesist.

Author details

Daniel Devote Basil[1*], Lawrence Hart[1] and Gbenga Ajayi[2]

1 Faculty of Environmental Sciences, Department of Surveying and Geomatics, Rivers State University, Port Harcourt, Nigeria

2 Faculty of Engineering and Built Environment, Department of Land and Spatial Sciences, Namibia University of Science and Technology, Namibia

*Address all correspondence to: basildevote@gmail.com

References

[1] Devi R. Oceanography: The Surface of the Sea. Ansari Road, New Delhi: Random Publications; 2018

[2] Bavestrello G, Silva N, editors. Oceanography and Benthic Ecology of Patagonian Fjords - 500 Years from the Discovery of the Strait Magellan. Lausanne: Frontiers Media SA; 2022. DOI:10.3389/978-2-88976-889-9

[3] Hill ED. From Seafloor Mapping to clean energy: hydrograph's key role in sustainable world. Hydro International Journal. 2023;**27**:3, 8-11

[4] Xu G, editor. Sciences of Geodesy. Vol. 1. Berlin Heidelberg, Dordrecht London, New York: Springer-Verlag; 2010 ISBN 978-3-642-11740-4

[5] Santosh KG. Ecology and Environmental Studies. Nai Sarak, Delhi: Khanna Publishers; 2006. ISBN: 81-7409-218-8

[6] IHO. Manual on Hydrography (M-13) - First Edition May 2005. International Hydrographic Bureau MONACO; Cedex Principauté de Monaco. 2005

[7] Ojinnaka O. Principles of Hydrographic Surveying from Sextant to Satellite. Enugu, Nigeria: El' Demark Publishers; 2007

[8] NASA. Seafloor Features Are Revealed by the Gravity Field. 2016. Available from the NASA Visible Earth website: http://visibleearth.nasa.gov/

[9] Torge, Muller. Geodesy. 4th ed. Berlin: Walter de Gruyter; 2012

[10] Lu Z, Qu Y, Qiao S. Geodesy: Introduction to Geodetic Datum and Geodetic Systems. Berlin Heidelberg: Springer Publishing; 2014 ISBN:978-3-642-41244-8

[11] Menandro PS, Bastos AC. Seabed mapping: A brief history from meaningful words. Geosciences. 2020;**10**(7):273

[12] Paolo T, Simon MM, editors. Remote Sensing for Geomorphology. Developments in Earth Surface Processes. Vol. 23. AE Amsterdam, Netherlands: Elsevier; 2020

[13] Abraham D. Modern Acoustics and Signal Processing. Switzerland: Springer Nature; 2019. ISBN: 978-3-319-92981-1

[14] Ferreira IO, Andrade LCD, Teixeira VG, Santos FCM. State of art of bathymetric surveys. Bulletin of Geodetic Sciences. 2022;**28**(1):e2022002

[15] Basil DD, Hart L, Tamuno JPK. Assessment of the effect of sensor misalignment of a multi beam hydrographic survey. World Journal of Geomatics and Geosciences. 2022;**2**:1

[16] Blondel P. The Handbook of Sidescan Sonar. Spring Praxis Publishing Ltd.; Berlin, Heidelberg: Springer; 2009. ISBN: 978-3-642-43463-1

[17] Urick RJ. Principle of Underwater Sound. Ontario: McGraw-Hill Ryerson, Ltd; 1983

[18] Finkl CW, Andrew JL. Shelf geomorphology along the Southeast Florida Atlantic continental platform: Barrier's coral reefs, nearshore bedrock, and morphosedimentary features. Journal of Costal Research. 2008;**24**(4):823-849. DOI: 10.2112/08A-0001.1

[19] Lurton X. An introduction to underwater acoustics. In: Lurton X, editor. Principles and Applications. Springer; 2002. p. 347. ISBN: 3-540-429670

[20] Makowski C, Finkl CW. Seafloor Mapping along Continental Shelves. London: Springer; 2016. ISBN: 978-3-319-25119-6

[21] Manik HM, Rohman S, Hartoyo D. Underwater multiple objects detection and tracking using multibeam and side scan sonar. International Journal of Applied Information System. 2014;**2**(2):1-4

[22] Lubis MZ, Anurogo W, Chayati SN, Sari LR, Taki HM, Pujiyati S. Sonar investigation and marine seismic investigation. Journal of Physics; Conference Sciences Series. 2020;**1442**. DOI: 10.1088/1742-6596/1442/1/012004

[23] Godin A. Calibration of Shallow Water Multibeam Echo-Sounding Systems. M. Eng. Report, Department of Geodesy and Geomatics Engineering Technical Report No. 190, Fredericton, Canada: University of New Brunswick; 1998. 182 pp